Modern Automotive Electrical Systems

Scrivener Publishing
100 Cummings Center, Suite 541J
Beverly, MA 01915-6106

Publishers at Scrivener
Martin Scrivener (martin@scrivenerpublishing.com)
Phillip Carmical (pcarmical@scrivenerpublishing.com)

Modern Automotive Electrical Systems

Edited by
Pedram Asef
Sanjeevikumar Padmanaban
and
Andrew Lapthorn

WILEY

This edition first published 2023 by John Wiley & Sons, Inc., 111 River Street, Hoboken, NJ 07030, USA and Scrivener Publishing LLC, 100 Cummings Center, Suite 541J, Beverly, MA 01915, USA

For more information about Scrivener publications please visit www.scrivenerpublishing.com.

Wiley Global Headquarters
111 River Street, Hoboken, NJ 07030, USA

For details of our global editorial offices, customer services, and more information about Wiley products visit us at www.wiley.com.

Library of Congress Cataloging-in-Publication Data

ISBN 9781119801047

Cover image: Electric Car Lithium Battery Pack | Erchog | Dreamstime.com
Cover design by Kris Hackerott

Set in size of 11pt and Minion Pro by Manila Typesetting Company, Makati, Philippines

Printed in the USA

10 9 8 7 6 5 4 3 2 1

Contents

1

General Introduction and Classification of Electrical Powertrains

Johannes J.H. Paulides*, Laurentiu Encica, Sebastiaan van der Molen and Bruno Ricardo Marques

Advanced Electromagnetic Group, Waalwijk, The Netherlands

Abstract
Did you know that all functionalities of our body are controlled by smart, super-intelligent, electrical systems? We actually do need these multi-billions of electrical signals in our nervous system to transfer the information via our central and autonomous nervous system to maintain and control, organs, muscle functions, and brain and spinal cord operation. The origin of these signals are actually electrochemical potentials that are perceptive in the nerve cells. The measurement of such electrical signals and potentials in the nerve transmission (similar to bodily responses) contains extremely valuable information of both the source as the receiver. Why start a powertrain book like this? Maybe it will become clear if we talk about an electrochemical battery and a receiving and returning powertrain? Anyhow, we wish you all the pleasure of reading/studying this book; it might be full of surprises and different points of view, but we trust that you will learn a thing or two.

Keywords: Electrical machines, electric drives, electric vehicles, micro-mobility, design process

1.1 Introduction

Electrical mobility is not only controlled using electrical systems, but supplied by electrical currents and sometimes even energy is firmly stored in permanent magnets, loosely stored in capacitors or temporarily

**Corresponding author*: info@ae-grp.nl

Pedram Asef, Sanjeevikumar Padmanaban, and Andrew Lapthorn (eds.) Modern Automotive Electrical Systems, (1–54) © 2023 Scrivener Publishing LLC

stored in batteries. To achieve the efficient control of electrical powertrains, we mainly use bytes grouped in open-CAN messages that are transmitted by electrical pulses. We also need multi-billions of these electrical signals to control the powertrain over its entire lifetime, as shown in Figure 1.1.

The sources of these communication electric signals are voltages (Joules) that, for example, find their electrochemical origin in batteries, and can be (temporarily) stored in capacitors. Measurements of these voltages and currents from the battery to the inverter and from the inverter to the motor provides extremely useful information about rotor position, temperatures, state of health, bearing deterioration, efficiency, inductance matrixes and other motor parameters, partial demagnetization, winding deterioration, cooling, winding resistance estimation, condition monitoring, etc.

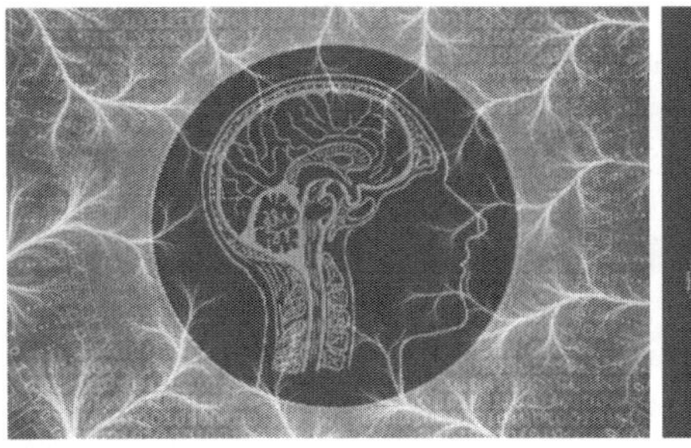

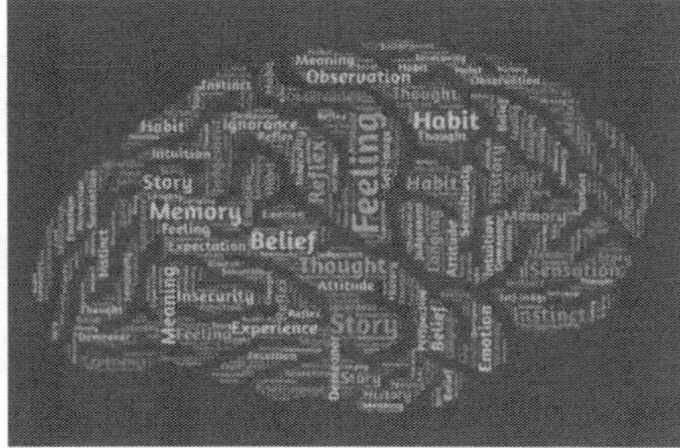

Figure 1.1 Multi-billions of electrical signals are transported through our nervous system [1].

Figure 1.2 Electrical pulses with an electrochemical origin? In this case maybe the Flintstone mobile should be considered as the first electrically propelled vehicle? [2].

In the context of electrical pulses that have an electrochemical energy storage origin, the Flintstone mobile might represent the first electrically propelled vehicle, as shown in Figure 1.2. In this respect, even walking could be classified as electrical mobility. However, a more common way is to classify mobility based on its electrochemical energy storage. As such, one of the earliest "controllable" energy storage device seems to be the "Leyden jars" that stored high-voltage electric charges (from an external source) between placed conductors on the outside and inside of a glass jar. Could this be considered a first in the sense that electrical energy became (kind of) mobile or could be deployed in a useful manner in an electrical network? Of course, batteries followed and this started electrical mobility. Today, electrical mobility, as shown in Figure 1.3, is something we take for granted on our roads, but the "road" to get where we are today was not so straightforward. Here we see the influence that we, as modest users with seemingly little influence, still have on getting a seemingly impossible-to-move and extremely powerful automotive industry to radically change! Although very few people thought it possible to alter the automotive industry, it did happen. We can agree or disagree on all the environmental influences of electrical mobility, but at the least it does remove pollution from distributed exhausts to an energy generation location and additionally allows for the "local use" of "local" energy generation sources. Only if such centralized or distributed energy generation sources become really sustainable will we all benefit. The main reason why electrical energy has proven to be a very useful form of energy is that it has the "potential" to be transported with almost 100% efficiency over relatively long distances. Meanwhile, what remains one of the challenges to be solved in the 21st century is creating (mobile) electrical energy storage, since electrical energy is

Figure 1.3 Electrical mobility is something we take for granted on our roads.

not easily stored and therefore mostly needs converting to another form of energy to allow bidirectional utilization.

We all know that transportation of people and goods will change very rapidly both in the short-term and long-term future. As such, green thinking, self-driving vehicles, and urbanization are just a few contemporary trends that will change transportation. Particularly important are the automotive industry business models to create solutions that will address current and future mobility trends. A vehicle with a 1980s user interface is not acceptable anymore for current customers that are used to mobile phones, interactive television and advanced domotica within their houses. Therefore, a shift from being a traditional car manufacturer with a business model focused on product sales toward being a provider of mobility services seems to be one promising approach. Furthermore, stricter legislative restrictions provide new challenges that will encourage (or even require) highly efficient and effective vehicle development solutions. New technologies, such as low energy loss tires, autonomous driving, vehicle driver interaction, the internet of things, and powertrain-specific technologies, provide a huge number of possibilities to optimize the way vehicles and the environment interact. Further, lifetime analyses and environmental considerations, as shown in Figure 1.3, will also be key enablers for sustainable mobility and influence the development, manufacturing, use, and recycling of vehicles.

This chapter provides a mobility systems overview and will try to identify their future challenges from both a contemporary, legislative and future perspective. The integration of electric components into the automotive powertrain enables completely different powertrain architectures and configurations to be designed. These components also enable additional

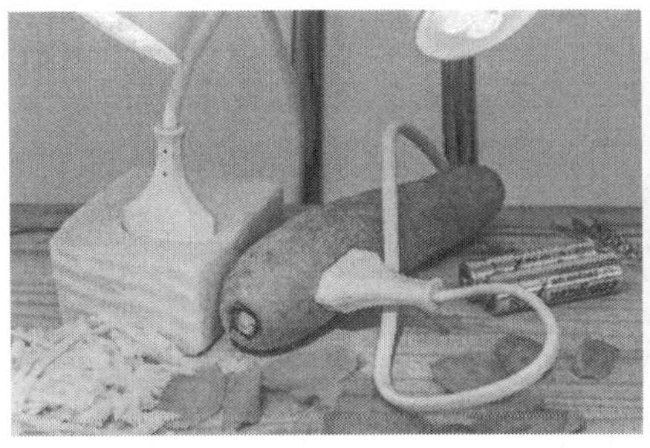

Figure 1.3 We need to think more about possibilities to optimize the way vehicles and the environment interact.

powertrain functions such as engine start-stop operation, silent powertrains, complete frictionless regenerative braking, electrical anti-lock braking, electrical differential and all these with new interfaces to the driver. Advanced or fully autonomous driver assistant systems are either under development or already on the market and require new interfaces to the vehicle and even new ways that vehicles might be used in the future, e.g., car sharing, vehicle robots, and driverless cars.

Future trends will lead to a higher diversification of powertrain systems with complex and altering software functionality, extendable or adjustable firmware and hardware, and advanced control or remote operation opportunities. This in turn results in higher powertrain software and commutation complexity and increased overall development effort. There is a need to know the most relevant powertrain functions, architectures and powertrain elements. Also, which development challenges are resulting from this powertrain system diversity? The automotive electrification provides rapid changes to powertrains, and such transformation brings engineering design challenges to consider integration, remote access and advanced control of these electrified elements. Powertrain-related targets are achieved by altering and harmonising various powertrain elements. Powertrain engineers will face new challenges in the design and manufacturing of batteries, e-drives, transmissions, low loss and long lifetime tires, body shapes, capacitors, contactless charging (even during driving), solar cells, and maybe fuel cells. System engineering will remain of utmost importance to take advantage of and understand all possibilities of (and in) electrified powertrains. The vehicle energy minimisation of subsystems as thermal management system, electrics/electronics (E/E), powertrain, body, chassis, or driving assistance will increase as interacting functions and (mechatronic) systems continue to increase, both in absolute numbers, as sales arguments. Therefore, we need to adjust, monitor, continuously coordinate and optimize development tasks across departments and (in the past considered impossible!) even beyond company boundaries. Managing the powertrain development complexity in term of organization and technology and considering its interfaces both upwards to vehicle development and downwards to the powertrain's elements and their manufacture will require a structured powertrain development processes that must be followed in the future accordingly. Simply focusing on individual elements or even subsystems and not considering the process and technical interfaces between them will prove insufficient. Most likely this is also why new vehicle manufacturers are currently so highly valued compared to traditional original equipment manufacturers (OEM). At this moment, numerous new companies are entering a market that was limited to a few

Figure 1.4 Powertrain complexity increases due to the growing number of interacting functions and (mechatronic) systems.

seemingly large automotive OEMs. In both these newly formed companies, as in more traditional OEMs that have been able to alter their course, the electrical powertrain software integration in future vehicles will remain a system engineering challenge, as shown in Figure 1.4.

This book provides insights into a powertrain development process for passenger vehicle systems. What is this powertrain system and consequently system engineering? For sure, it has a wide scope and interpretation. However, does it not just relate to a structured and connected way of thinking and working? Electrified powertrain "systems" will always be ever-more complex, mainly caused by the growing number of required functions to cover trends such as autonomous driving and cybersecurity, and still comply with ever-lagging legislation. This growing number of functions, that are fully recognized by the automotive customer, also causes an increase in integration effort. Four structuring principles and so-called systems engineering core development methods can be identified. Starting with the method to determine the requirements for the engineering, consequently with the system specifications, and then system integration, and finally the system verification and validation.

1.2 Worldwide Background for Change

Urban mobility across Africa, Asia, Oceania, South and North America, and Europe is witnessing ubiquitous growth, but ever-more micro-mobility in the form of shared electric bikes, scooters, and motorbikes will be available. Is it true that modern consumers really prefer micro-mobility? The convenient combination of electric traction and (future) dockless operation, and the improvements in human-machine interface, do allow for more intuitive use of these new forms of transport. Or is it mostly the younger/current generation that adopts micro-mobility since for them it is seamless to use a two-wheel vehicle via a smartphone app rather than a traditional transport system such as a car, bus, or train that is dependent on road-traffic/state? Are we not forgetting about a generation? Is it not so that the human-machine interface is more a "youth"-machine interface? And will the youth of today still feel comfortable with the mobility of the future, when we get flying cars, hence more three-dimensional transportation? Slowly, we all realise that we need to change how we treat this, but what is the best mobility to achieve that? For sure, it will not be a 3000kg electrical vehicle to move a person weighing around 80kg. Actually, it is pretty amazing that such heavy mobility even exists and although carmakers are getting used to heavier vehicles, and all structural changes that are associated with this increase in mass, it is certainly not the most efficient means of transportation. It was only a few years ago that every kilogram and carbon dioxide saved was the aim of all vehicle engineers and even financial bonuses were associated with such innovations. Very few at that time were anticipating the current trend concerning the extreme mass of vehicle today. For sure, future electrochemical devices will be lower in mass and a new race to the lightest mobility is starting now. In the future, the same will happen with three-dimensional or (maybe) even four-dimensional mobility.

Urbanization is happening throughout the world, although pandemic concerns and working at home might slow down the trends of extreme congestion in the city areas. Still, whatever will happen, we realize that time lost by passengers, congestion pollution, and space taken to provide current mobility in urbanized areas will not be accepted in the future, as shown in Figure 1.5. Roads are sometimes accounting for more than 20% of available city areas, and would it not be nice to have green parks and areas instead of roads?

Figure 1.5 Congestion-free future cities with energy-efficient people mobility concepts.

Electrical micro-mobility will positively impact air pollution and opportunities for roads to be converted into green parks. At this moment, air pollution is one of the largest struggles for cities across the world. Micro-mobility start-ups offering emission-free two-wheel vehicles are emerging like mushrooms and provide an effective alternative (or addition) to the existing private and public transit mix. Especially considering the current pandemic situation, but also more normal times in the near and distant future. Will people resurface after this pandemic and still select an open-air, single-rider transit vehicle that allows more naturally to maintain social distancing and so ensure a minimum of shared points of contacts? Besides the COVID-19 pandemic, we also have to take into consideration the rising global population, which will reach around 8.5 billion by 2030 [3]. Even more cumbersome is that it is estimated that around 60% of these people will be living in "more" urbanized areas, and they will account for about two-thirds of the total global emissions if we do not dramatically change the more traditional transportation to more environmentally friendly micro-mobility.

A further urbanized population will cause an exponential increase in congestion, the major contributor being traditional transportation vehicles such as car, tram, bus, or non-matching train schedules. On the one hand, it is anticipated that micro-mobility will be a very cost-effective answer. On the other hand, there are still many micro-mobility challenges, such as lack of regulations, citywide bans, charging, visual pollution, fire prevention and extinguishing, and theft. The e-bike batteries fire hazard is something that is being noticed more and more and poses an increased risk; thermal runaway remains problematic. What about measures for safe storage and handling, as shown in Figure 1.6? Some of these topics will be discussed in this chapter and the current legislative situation is provided in more detail.

Figure 1.6 Immersing an electric vehicle in water to extinguish the battery [4], and former e-bike battery [5].

Micro-mobility will grow, mostly due to the increasing smartphone adoption and due to the advantage of a need for first mile – last mile connectivity. The millennial generation is very familiar with smartphones along with GPS advancements, 24-hour connectivity, and mobile payments. All this will lead to lower commutation cost; for example, a typical bicycle-sharing service includes an initial cost of €0.50 and an additional €0.50 per 30 minutes. This is of course much less than other public transportation sharing options. In terms of funding, there also seems to be an improving environment for investments. The rising micro-mobility adoption in Europe, Asia, Oceania, Africa, and South and North America will also help to increase the market potential. As such, micro-mobility service has been introduced in nearly all geographies globally, and Asia and North America have witnessed extremely significant growth. Especially in Asia, where nearly half of all citizens live in urban areas. The continent's cities are expanding at an unprecedented rate, therefore road congestion and air pollution need to be reduced. In this respect, sustainable urbanization will be extremely important, where inhabitants will be a deciding factor in micro-mobility vehicle adaptation. It also seems that dockless facilities offered by private companies contribute to a decrease in the use of "more" public bike-sharing. Although numerous companies are offering locally, international expansion is still a challenge, since it is difficult to manage high operational costs and competition. In North America, dockless electric scooters will be more implemented compared to e-bikes. Apart from start-ups, North America's larger ride-hailing companies are starting to introduce all forms of transportation into their portfolio, including larger mobility platforms, option to travel by train, vehicle sharing, etc., as shown in Figure 1.7.

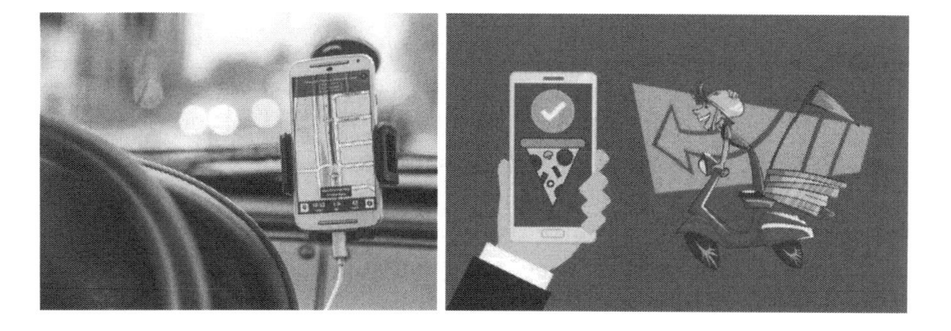

Figure 1.7 Navigation to food/ride-hailing companies are introducing all forms of transportation.

Why is electric mobility, which also allows various forms of higher-level vehicle platforms with tracking and online connections, suddenly becoming so interesting? Currently all automakers are either selling or developing "zero-emission vehicles" with "zero-emission supply chain", where the momentum behind cleaner, software-enabled electrical mobility forms is seemingly unstoppable. In the past, human-vehicle interaction was at a low level and still the main vehicle feature was to get from A to B. The widespread introduction of navigation, followed by in-vehicle communication and electrical powertrain, changed this dramatically. The amount of available computational power to operate an electrical powertrain is much higher compared to conventional powertrains. Therefore, much more electronic components and software will be needed in vehicles. This immediately introduced new competencies that were required at vehicle OEMs. Over the next five to ten years, the skill sets of traditional automakers and suppliers will need to shift dramatically. As we all are aware, the rising importance of in-vehicle software due to connected and/or autonomous cars pushes emphasizes the need of digital talent. Nowadays, vehicle OEMs are (or will be) in direct competition with digital giants, for example ASML in the Eindhoven region in the Netherlands. However, it is not only about seeking new kinds of engineers; a completely different environment will be present from the reduced (or complete lack of) vehicle maintenance and the switch to online instead of showroom sales. This will affect the entire automotive value chain, as consumers are ever more comfortable with online purchases, hence local dealers will be eliminated. A "forgotten" advantage of an electrical powertrain is that it also significantly reduces the number of components and hence the assembly time. In this respect, both factory workers and current powertrain suppliers will be affected by this shift to electrical. Electrical powertrains require engineers with electrical and electrochemical skills. Further, companies will have to become more skilled at using digital tools and we will see the introduction of shared service centers in lower-cost locations. An extra focus will be on software and data analytics as they are essential for online sales and marketing. Another "forgotten" item is the need for a culture of transformation, where vehicle OEMs will have to provide other employee experiences that top talent has come to expect at tech companies. This means approachable managers, speedy hiring and onboarding, agile-work and fast-decision processes with quarterly business reviews and objectives, key results methodologies, compensation and incentives tied to talent rather than hierarchy, and an acceptance of flexible/remote work, as shown in Figure 1.8.

Beyond, industry 4.0 requests a workforce shift or even a culture transformation and this is something that is not easy for existing OEMs, hence

Figure 1.8 The required culture transformation is something that is not easy for exiting OEMs.

new OEMs are popping up as culture is not something that can be changed overnight. These new players are supported by the capital market and are outperforming the traditional automotive industry. It seems that new mobility start-ups are accessing new value pools. Public attention is mainly on autonomous electrical vehicles; however, the mobility platforms head in several other directions. We might have to get used to personalized contextual advertising using advanced connectivity solutions that are available in electrical vehicles. This will result in a considerable new, currently untapped, subscription potential. All grids, mainly due to the combination of centralized and decentralized generation, are slowly becoming more "green"; hence the discussions of where our electrical power is coming from is also relevant. These new mobility platforms value zero-emission vehicles, where consumers are increasingly factoring sustainability into their buying decisions. Further, instead of standardized high volume vehicle production, the focus is more on customer experience. Customer-centric vehicles with on-board robotics and human interaction, e.g., Siri, Alexa, and Nomi, will outperform standardized platforms, and car buying and maintenance will become experience shopping that will require complete change in consumer journey. The current showroom will not be a mere website nor dedicated apps, but blogging, social networks, flagship showrooms, and vehicle owners' lounges, actually creating a pleasing environment to guide the consumers from vehicle sales to upgrade packages, consumer-oriented alterations, connectivity and current and future service upgrades. Please note that the absence of the traditional dealer saves up to 25% on each vehicle. These new mobility sales, design, development and production requires new kinds of talent. OEMs will be software- and/or electronics-first companies and as such will have full focus on hiring digital talents. More traditional OEMs have to reskill many of their current workforce. Western electric-car owners will have a different culture from, for example, Asian owners. Here, the vehicle inside and outside will need to be merged. These new Asian-based vehicles need to deliver a friendly and convenient vehicle, starting with online sales. When entering the vehicle, advanced facial-recognition systems are used to adjust seat, steering wheel, and interior lighting. The dashboard screen will welcome the driver and will offer restaurant recommendations in case they have an appetite; changing the music by voice control, and parking and charging will be automatically arranged during work hours. The dashboard also allows the driver to pay for charging, coffee, tea and even restaurant bills. The Asian consumer is willing to pay about twice as much for such features. The differentiating feature of future vehicles will be their ability to mimic smartphone conveniences.

1.3 Influence of Electric Vehicles on Climate Change

Are the current European climate targets sufficient to meaningfully limit climate change? One thing is sure, minimizing carbon emissions from vehicles is extremely critical. For example, transport represents nearly a quarter of Europe's greenhouse gas emissions, but maybe more importantly, it is the main cause of air pollution in cities and other highly urbanized areas. In this respect the transport sector has not (yet) seen the same gradual decline in emissions as other sectors have: emissions only started to decrease in 2007 and still remain higher than in 1990. Within this sector, road transport is by far the biggest emitter, accounting for more than 70% of all GHG emissions from transport in 2014 [6]. Nobody envisaged that electrochemical technology development within the mobility is proceeding at a very fast pace. For example, batteries are now around $100 per kilowatt-hour, down from $1,200 in 2010, where electric vehicle–related technologies investment and assets almost doubled in 2020. And yet . . . it is not enough for the 1.5°C climate targets, as shown in Figure 1.9.

The decade between 2025 and 2035 will determine whether the industry can keep cumulative CO_2 emissions for passenger cars (through 2050) to under 45 gigatons, a "carbon budget" that would help hold global temperature increases to under 1.5°C [7]. Mobility players, consumers, and governments need to increase the large-scale adaptation of zero-emission vehicles. This is the most effective way to reduce carbon emissions. Everybody should realize that by 2035, more than 95% of all cars and trucks on the road must be zero-emission to limit global warming to 1.5°C. Of course, hybrids, increased share of renewables, travel distance reduction, and zero-carbon supply chains also play a role. For the complete vehicle, it is

Figure 1.9 Electric mobility and the 1.5°C climate target.

of utmost importance that materials, material transportation, and in-plant processes should also become zero-emission. This task is huge! It requires a complete cross-industry effort. Ultimately, regulators and policy makers have to intercede, adopting an ecosystem perspective with cross-border goals and meaningful timelines for everything from number of charge points needed to carbon reduction. Without this, it will be unlikely that the shift to electric vehicles will be realized to its full potential for creating a sustainable value chain and for mitigating the impact of climate change. The means by which governments can influence the road-legal vehicles is by defining mobility classes and influencing the vehicle mix by mobility incentives. This has been deemed so important for future engineers that the considerations have been included in this chapter. As researchers we might be biased toward focusing completely on the technical aspect, but if we forget about the regulations, we might design or develop an electrical powertrain that will be very difficult to be road legal.

1.4 Mobility Class Based on Experience in the Netherlands (Based on EU Model)

Type-approval within the European Union (EU) is allowed in one member state only but that type-approval is not (yet) valid throughout all member states of the European Union [8]. The national approval authority officially certifies that a component, separate technical unit, vehicle, or system, is approved by means of the *type-approval* certificate. Following this, it is up to the manufacturer to issue a "certificate of conformity", i.e., a certifying document to "prove" that the produced vehicle/system/component/ conforms to the approved product and this needs to be signed by the CEO of the company. For all parties in the procedure, the type-approval

Figure 1.10 Micro-mobility – last mile.

legislation lists the obligations. Manufacturers need to ensure that their products are manufactured in accordance with type-approval requirements. Manufacturers must also ensure that procedures are applicable for series production to retain conformity with the approved type. In case of non-conformity or serious risks appearing from certain products, manufacturers have specific obligations to fulfil. For example, if they are established outside the EU, they must appoint a single representative in the EU before the type-approval authority.

As for dealers, they have the following obligations [9]:

- Verify that product bears the required statutory marking or type-approved mark.
- Verify that product is accompanied by the legally required documents and safety information in the official language(s) of the member state.
- Verify that product is accompanied by the certificate of conformity.
- Verify that name, registered trade name or registered trade mark and the address at which the manufacturer can be contacted is on the vehicle or on packaging or in a document with the vehicle. Check that the required statutory plate with the appropriate marking is affixed.
- Check that each component has the required type-approval mark. If a component does not need type-approval, the manufacturer must at least affix a trade name or a trade mark and a type number or identification number.
- In case of non-conformity or serious risk the dealer must inform the manufacturer.
- Ensure that type-approved components are replaced only by type-approved components.

Please note that in the Netherlands we have the following mobility categories:

L Motor vehicles with less than four wheels [but does include light four-wheelers]

L1 A two-wheeled vehicle with an engine cylinder capacity in the case of a thermic engine not exceeding 50 cm^3 and whatever the means of propulsion a maximum design speed not exceeding 50 km/h. (Electric bicycle)

L2 A three-wheeled vehicle of any wheel arrangement with an cylinder capacity in the case of a thermic engine not exceeding 50 cm^3 and whatever the means of propulsion a maximum design speed not exceeding 50 km/h. (Auto rickshaw)

L3 A two-wheeled vehicle with an engine cylinder in the case of a thermic engine exceeding 50 cm3 or whatever the means of propulsion a maximum design speed exceeding 50 km/h. (Motorcycle)

L4 A vehicle with three wheels asymmetrically arranged in relation to the longitudinal median plane with an engine cylinder capacity in the case of a thermic engine exceeding 50 cm^3 or whatever the means of propulsion a maximum design speed exceeding 50 km/h (motor cycles with sidecars).

L5 A vehicle with three wheels symmetrically arranged in relation to the longitudinal median plane with an engine cylinder capacity in the case of a thermic engine exceeding 50 cm^3 or whatever the means of propulsion a maximum design speed exceeding 50 km/h. (Motorized tricycle)

L6 A vehicle with four wheels whose unladen mass is not more than 350 kg, not including the mass of the batteries in case of electric vehicles, whose maximum design speed is not more than 45 km/h, and whose engine cylinder capacity does not exceed 50 cm^3 for spark (positive) ignition engines, or whose maximum net power output does not exceed 4 kW in the case of other internal combustion engines, or whose maximum continuous rated power does not exceed 4 kW in the case of electric engines. (Golf cart, Mobility scooter)

L7 A vehicle with four wheels, other than that classified for the category L6, whose unladen mass is not more than 400 kg (550 kg for vehicles intended for carrying goods), not including the mass of batteries in the case of electric vehicles and whose maximum continuous rated power does not exceed 15 kW. (Microcars)

M Vehicles having at least four wheels and used for the carriage of passengers (e.g., standard car with 2, 3, 4 door).

M1 Vehicles used for carriage of passengers, comprising not more than eight seats in addition to the driver's = 9. (Larger Than Standard Car e.g.: London Cab/E7 Type Vehicle 8 seat +Driver.)

M2 Vehicles used for carriage of passenger, comprising not more than eight seats in addition to the driver's seat, and having a maximum mass not exceeding 5 tonnes. (Bus)

M3 Vehicles used for the carriage of passengers, comprising more than eight seats in addition to the driver's seat, and having a maximum mass exceeding 5 tonnes. (Bus)

N Power-driven vehicles having at least four wheels and used for the carriage of goods

N1 Vehicles used for the carriage of goods and having a maximum mass not exceeding 3.5 tonnes. (Pick-up Truck, Van)

N2 Vehicles used for the carriage of goods and having a maximum mass exceeding 3.5 tonnes but not exceeding 12 tonnes. (Commercial Truck)

N3 Vehicles used for the carriage of goods and having a maximum mass exceeding 12 tonnes. (Commercial Truck)

O Trailers (including semi-trailers)

O1 Trailers with a maximum mass not exceeding 0.75 tonnes.

O2 Trailers with a maximum mass exceeding 0.75 tonnes, but not exceeding 3.5 tonnes.

O3 Trailers with a maximum mass exceeding 3.5 tonnes, but not exceeding 10 tonnes.

O4 Trailers with a maximum mass exceeding 10 tonnes.
Special purpose vehicle [Cat 1]

Motor caravan, [Cat 2] also Campervan, Motorhome.
Armoured car [VIP], Armoured car (valuables) [Cat 3]
Ambulance [Cat 4]
Hearse [Cat 5]

T Agricultural and Forestry tractors [Cat 6]
Non-road mobile machinery [Cat 7]

G Off-road vehicles [Cat 8] [Cat 9]
Special purpose vehicles (M1)

Example Micro-Mobility
If we take the example of micro-mobility, e.g., "electric bicycles", we have to realize that all e-bikes (or pedelecs) are subject to type-approval with the exception of the following:

- Vehicles exclusively intended for use by the physically handicapped (for instance three-wheelers, hand-bikes, etc.)
- Vehicles exclusively intended for use in competition
- Vehicles designed and constructed for use by armed services, civil defence, fire services, forces for monitoring public order and emergency medical services
- Vehicles primarily intended for off-road use and designed to travel on unpaved surfaces
- Pedal cycles with pedal assistance which are equipped with an auxiliary electric motor having a maximum continuous rated power of less than or equal to 250 W, where motor output is cut off when pedalling stops or is progressively reduced and cut off before the vehicle speed reaches 25 km/h
- Vehicles with an R-point height of < 540 mm for L1e (for instance recumbent bikes, velomobiles with a saddle below 54 cm)
- Self-balancing vehicles (for instance Segway) and
- Vehicles with not one seating position (for instance Trikke, Egret, etc.).

Consequently, manufacturers around the world are confronted with very divergent requirements. This is why, within CEN TC 354, a standardisation procedure has been initiated. So, most likely in a few years' time, self-balancing vehicles and vehicles with no seating position will become subject to one European harmonised technical standard.

1.5 Type-Approval Procedure

The type-approval legislation is made up of the framework Regulation 168/2013, which lays down the basis of the type-approval. The competence for this law was with the European Parliament and Council. All administrative and technical details are from the European Commission, which laid down the following four Regulations:

- Delegated regulation on functional safety
- Delegated regulation on vehicle construction
- Delegated regulation on propulsion unit and environmental performance
- Implementing regulation on administrative provisions.

Previously, legal texts had to go through Parliament and Council, which made it a very time-consuming procedure. In the new procedure, only the basic text and administrative and technical details are undertaken by the Commission. This ensures that it is easier and quicker to amend and correct details if necessary and/or adapt them according to technical progress. Originally, draft legal texts were written exclusively in regard to type-approving conventional mopeds and motorcycles. Yet the scope of the type-approval now also includes most electric bicycles. ETRA, the former dealer association, first drew the European institutions' attention to the fact that technical requirements for mopeds and motorcycles are not necessarily best suited for electric bicycles. Eventually, the European Commission agreed to cooperate with the electric bicycle business and together they adapted the electric bicycles type-approval requirements. Later, COLIBI/COLIPED also joined, which resulted in a type-approval system that to, a certain degree, is adapted to electric bicycles.

Figure 1.11 E-bikes that are limited to 250W [10].

The system is of course not perfect yet, so in the next years, further improving and updating will be required. The component list and characteristics subject to type-approval is the framework Regulation 168/2013. Some micro-mobility electric bicycles applications are excluded from type-approval for some components and characteristics, whereas for others specific requirements have been introduced. Those that do not feature in the list are excluded from electric bike type-approval. The type-approval must be carried out by a technical service, which has been designated by the approval authority of a member state, as a testing laboratory to carry out tests or as a conformity assessment body. The approval authority of a member state is established or appointed by the member state and notified to the European Commission. The approval authority is competent for all aspects of type-approval, i.e., issuing, withdrawing or refusing approval certificates.

Every micro-mobility vehicle has to come with marking and instructions as listed in the standards.

- the frame must be visibly and permanently marked with a serial number at a readily visible location
- the frame must be visibly and durably marked, with the name of the manufacturer or the manufacturer's representative and the number of European Standard, i.e., EN 14764
- the vehicle must be durably marked with the following words: EPAC according to EN 15194
- 25 km/h, i.e., cut-off speed
- 250 W, i.e., electric motor maximum continuous rated power (Figure 1.11).

Not only is type approval a requirement, but also compliance to the Machinery directive, for example, "Cycles with pedal assistance, equipped with an auxiliary electric motor with a 250W maximum continuous rated power, of which the output is progressively reduced and finally cut off as the vehicle reaches a speed of 25 km/h, or sooner, if the cyclist stops pedalling" are explicitly included in the scope of Directive 2006/42/EC on Machinery. This Directive contains a list of essential health and safety requirements related to the design and construction of machinery, i.e., micro-mobility vehicles. Vehicles may only be placed on the market and/or put into service if they comply with these requirements. Most of those health and safety requirements are covered by EN 15194. However, in the European standardisation institute CEN, TC 333 Micro-mobility "Cycles" is in the process of reviewing EN 15194 to ensure that all obligations resulting from the Machinery Directive are covered by the standard. Furthermore, whereas EN 15194 currently only covers the electric part of the bicycles and EN 14764 is applied to the mechanical part, the new standard will have technical requirements for the whole vehicle. The Machinery Directive holds a few additional administrative obligations for the manufacturers. They must have a complete technical file on the product available. Furthermore, they have to supply the micro-mobility vehicles with an EC Declaration of Conformity, the particulars of which are specified in Annex II of the Directive. The vehicle must have a CE conformity marking with the initials "CE". The CE marking shall be affixed to the micro-mobility vehicle visibly, legibly and indelibly in the immediate vicinity of the name of the manufacturer or its authorised representative. This marking, however, can only be affixed if the micro-mobility vehicle also conforms to Directive 2004/108/EC relating to electromagnetic compatibility and to the RoHS Directive. Please note that it is prohibited to pre-date or post-date the micro-mobility vehicles when affixing the CE marking.

Finally, in addition to the CE marking, the micro-mobility vehicle must be marked visibly, legibly and indelibly with the following minimum particulars:

- the business name and full address of the manufacturer and, where applicable, its authorised representative
- designation of the micro-mobility vehicle
- designation of series or type
- serial number, if any
- the year of construction, that is the year in which the manufacturing process was completed.

All electric devices influence each other when interconnected or close to each other. Sometimes one may observe interference between a TV set, a mobile phone, a radio and a nearby washing machine or electrical power lines. The purpose of electromagnetic emissivity (EME) is to keep all those side effects under reasonable control. Micro-mobility vehicles with an electric motor having a maximum continuous rated power of 250W and assisting up to maximum 25 km/h must comply with this Directive. The manufacturer has to apply its methodology for the EME assessment and has to prepare technical documentation to demonstrate evidence of compliance with the requirements and have that documentation available.

Further, the RoHS directive is applicable to not contain any lead, mercury, cadmium, hexavalent chromium, polybrominated biphenyls (PBB) or polybrominated diphenyl ethers (PBDE). The manufacturer has to certify that its product complies with the RoHS Directive by means of a CE marking. This, however, cannot be done unless the product also complies with the Machinery Directive and the EME Directive. In order to comply with the RoHS Directive, the manufacturer also has to draw up technical documentation, carry out an internal production control procedure and provide for a declaration of conformity.

Finally, battery transportation due to short-circuit of the battery as a result of the battery terminals coming into contact with other batteries, metal objects or conductive surfaces, is one of the major risks associated with the transport of batteries and battery-powered equipment. Therefore, their transport is subject to very strict rules, which have been internationally harmonised. Any Lithium-Ion battery over 100Wh is classified as CLASS 9 - MISCELLANEOUS DANGEROUS GOODS under the dangerous good regulations for transport by road (ADR), by air (IATA & IACO) and by sea (IMDG). The UN number for Lithium-Ion batteries is 3480, if contained in or packed with equipment 3481. These rules do not only concern transport of batteries, for instance, from manufacturer to dealer, but all transport, except transport for private purposes. Firstly, to ship goods in the CLASS 9 category means that the battery needs to be tested in accordance with the UN Manual of tests and criteria, Part III, subsection 38.3.1. Also, very specific and strict procedures related to handling, packing, labelling and shipping need to be followed. If any company handles, packs and labels dangerous goods, such as Lithium-Ion batteries at their own premises, a trained "Dangerous Goods Advisor" is required on-site to oversee that this is done in full compliance with the rules and to declare the goods safe to travel. If you have no member of staff which has received the above training, you must hire a specialist company to handle, pack and label the goods and to fill out a "Dangerous Goods Note". It is compulsory for Dangerous Goods shipments to be accompanied by this document. There are also weight restrictions for the transport of batteries. A package shipped by air containing a lithium battery may not exceed 10kg gross. The weight limit

Figure 1.12 Specific measures apply to industrial batteries [11].

per package shipped by road or sea is 30kg gross. There will be occasions where a manufacturer may wish to have a defective or damaged battery returned for analysis. However, such batteries are prohibited from transport by air. This prohibition also applies to waste batteries and batteries being shipped for recycling or disposal. The rules for transport of defective, damaged and waste batteries by road or by sea are inconclusive and are currently being discussed in the relevant international committees.

The following specific measures apply to industrial batteries:

- Producers must be registered in the national register of all Member States where they place batteries on the market for the first time. If, for instance, the manufacturer of the battery in an electric bicycle or the manufacturer of the electric bicycle or its representative are not registered nationally, the dealer will be considered to be the producer of the battery and will be held responsible.
- Producers of industrial batteries or third parties acting on their behalf have an obligation to take back waste industrial batteries.
- Industrial batteries have to be readily removable from electric bicycles. If the battery is integrated in the bicycle, it has to be accompanied by instructions showing how the batteries can be safely removed and who is the best person to do this.
- Batteries must be labelled with a crossed out wheeled bin and chemical symbols indicating the heavy metal content of the battery.
- All collected industrial batteries must be recycled. Industrial batteries may not be disposed of in landfills or by incineration.

Please note that the above list is not exhaustive and everybody should ask the national collection scheme for instructions (Figure 1.12).

As mentioned, there are several engineering challenges involved in the design of powertrains for mobility applications. Any powertrain has to account for limitations in volume and mass. Further, electrical powertrains are expected to operate continuously for years and be able to cope with wide temperature variations. In general powertrains we can distinguish two main operation regions, namely constant torque and constant power that "fit" the mobility application requirement. In both regions, the electrical current should be measured and adjusted to achieve the desired torque at the highest efficiency in forward motion. To maximise the torque to the wheels, the current reference and angle needs to be determined. Most of the times in mobility and considering the loading condition of the powertrain, this current reference and angle is set different from "aligned" in a process called field weakening. This is necessary when the motor is limited in volume and mass, as is the case in any powertrain, when the motor is intended to operate at speeds higher than the nominal speed or in case maximizing the torque requires a non-aligned current waveform with the synchronous motor back-emf waveform. The induced voltage, back electromotive force, is dependent on rotational speed and loading to utilize both reluctance and permanent magnet torque. In summary, all mobility classes and vehicles should consider field weakening to minimize motor volume and mass. However, what is field weakening?

1.6 Torque-Speed Characteristic of the Powertrain for Mobility Vehicles

In order to cover the claims made in the theory, an overview will be given about the assumptions on which this chapter is based. Based on these assumptions, the theory outlined should be correct.

- Prior to this document, the reader is assumed to have basic electrical machine control knowledge such as field-oriented control including the pseudo d-q transformation.
- The quantities and the theory of magnetism and electronics are assumed to be known.
- The use of phasor diagram, i.e., representation of permanent magnet synchronous motor model in d and q-axis reference frame, is based on the requirement of the machine to be sinewound. Meaning a machine that has a sinusoidal EMF waveform and a sinusoidal variation of inductance with rotor position [12].
- Motor parameter identification is undertaken correctly. This is not part of this chapter.
- A star connected permanent magnet synchronous motor with or without saliency is considered, not a hybrid (combination of reluctance and permanent magnets) nor rotor fields (using rotor field windings) can be altered and magnets cannot be remagnetised *in situ* in this chapter.
- In the convention adopted in this chapter, a negative value of Id implies that the armature flux opposes the PM field, and a positive value is indicative of aiding PM and armature fields. Likewise, a positive and negative I_q is

representative of operation in the motoring and generating modes, respectively.

- A number of requirements have to be balanced (torque/power, current and voltage limits, weight) and setting a list of priorities from a vehicle point of view will always be necessary.
- No information about the cooling system is given, and, therefore, a thermal analysis is not considered.
- The requirements for the motor operating envelope should be based on the expected load curve and how this should be covered within the given current and voltage limits (more details later in this chapter).

The Lotus Elise is a two-seat, rear-wheel drive, mid-engined roadster conceived in early 1994 and released in September 1996 by the British car manufacturer Lotus Cars. The Elise has a fibreglass body shell atop its bonded extruded aluminium chassis that provides a rigid platform for the suspension, while keeping weight and production costs to a minimum. It is capable of speeds up to 200 km/h (125 mph). The Elise was named after Elisa Artioli, the granddaughter of Romano Artioli, who was chairman of Lotus and Bugatti at the time of the car's launch. For this initial derivation

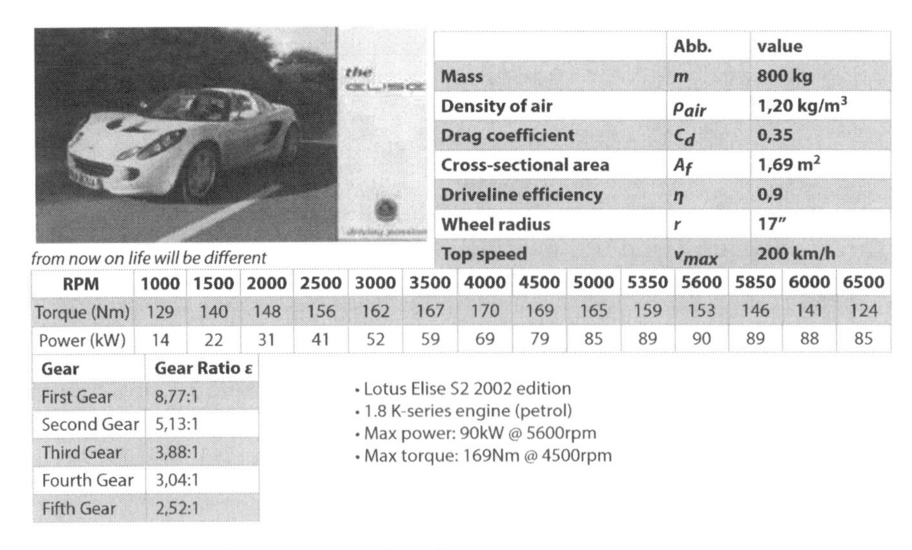

	Abb.	value
Mass	m	800 kg
Density of air	ρ_{air}	1,20 kg/m^3
Drag coefficient	C_d	0,35
Cross-sectional area	A_f	1,69 m^2
Driveline efficiency	η	0,9
Wheel radius	r	17"
Top speed	v_{max}	200 km/h

RPM	1000	1500	2000	2500	3000	3500	4000	4500	5000	5350	5600	5850	6000	6500
Torque (Nm)	129	140	148	156	162	167	170	169	165	159	153	146	141	124
Power (kW)	14	22	31	41	52	59	69	79	85	89	90	89	88	85

Gear	Gear Ratio ε
First Gear	8,77:1
Second Gear	5,13:1
Third Gear	3,88:1
Fourth Gear	3,04:1
Fifth Gear	2,52:1

- Lotus Elise S2 2002 edition
- 1.8 K-series engine (petrol)
- Max power: 90kW @ 5600rpm
- Max torque: 169Nm @ 4500rpm

Figure 1.13 Lotus Elise 2002 internal combustion engine parameters.

of the torque-speed curve, we will consider the Lotus Elise S2 2002 edition (Figure 1.13).

Considering the above, an analysis is made of the torque-speed characteristic of this Lotus Elise. It can be seen that a constant torque to a certain speed followed by an ever-decreasing torque is something that is acceptable for vehicles. Driving a 30% slope with 200kph is not required, hence it is accepted that a vehicle does less speed uphill. Actually, a second curve is shown in this torque-speed characteristic, where the electric motor power is limited to 50kW (Figure 1.14).

Considering this, reduced power has been undertaken to compare the forward motion characteristic of this vehicle as shown in Figure 1.15.

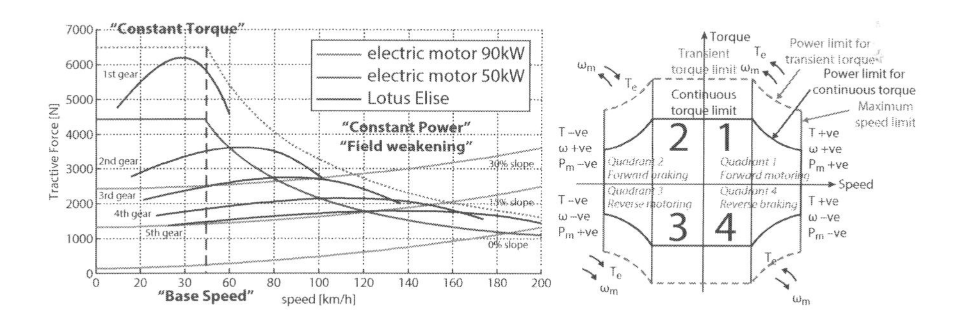

Figure 1.14 Torque-speed characteristics of a 2002 Lotus Elise and a more general torque-speed characterictics showing all four quadrants.

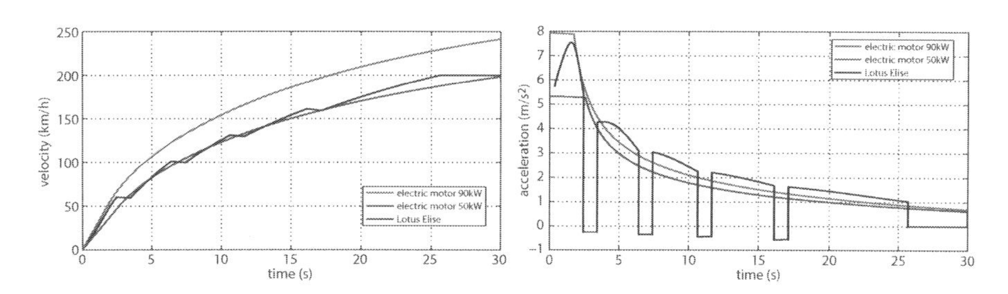

Figure 1.15 Shaft power of internal combustion engine Lotus Elise to an equivalent Elise with an electric motor (comparison 'only' covers the forward motion characteristic of the vehicle).

This clearly shows that actually 50kW at the shaft of the electrical machine is sufficient to match longitudinal vehicle characteristic of the 2002 Lotus Elise. This is also considered in modern-day electric vehicles, which not only try to minimize the drag coefficient, but also the nominal electric power. Actually, an electric motor is very special since it can be overloaded many times, as long as the thermal limits are not exceeded.

```
%% Engine parameters Elise
RPM =  [1000, 1500, 2000, 2500, 3000, 3500, 4000, 4500, 5000,
   5350, 5600, 5850, 6000  6500 ];
Torque_e – [129.56, 139.69, 145.87, 158.23, 166.23, 161.98, 165.52, 176.25,
   163.19, 158.13, 153.47, 145.87, 140.60, 124.30 ];
p = polyfit(RPM/1000,Torque_e,5);
Torque_e = polyval(p,RPM/1000);

% Electric motor parameters
Pelpeak = 90e3;   %[kW]
Pelnom = 50e3;   %[kW]
% Resistance parameters
m = 800;        %[kg] weight of the car
rho = 1.2041;  %[kg/m^3] density of air
Cd = .35;       % drag-coefficient
Af = 1.69;       % cross-sectional area

theta0 = 0;theta15 = 8.53;theta30 = 16.7;   % [deg] slope of the road => 10% =
   5.71, 20% = 11.3, 30% = 16.7

V = [0:200]/3.6; % velocity of the car
eta = .9;      % driveline efficiency
r = 17*2.54e-2/2; % wheelradius [m]

%% calculations % In gear speeds
GearRatio = [8.7707;5.1344;3.8809;3.0447;2.5183];
Gearspeed = repmat(RPM,5,1)./
   repmat(GearRatio,1,length(RPM))*2*pi*r/1000*60;
% Torques normalised by gear ratio
Torque = GearRatio*Torque_e*eta;
% Engine generated tractive effort
Force = Torque/r;

% Resistance calculations
Ra = rho/2*Cd*Af*V.^2;  %[N] aerodynamic resistance
frl = 0.01*(1+V/147);  %[0.01-0.015]
```

```
Rrl = frl*m*9.81*2;        %[N] rolling resistance
Rg0 = m*9.81*sind(theta0); Rg15 = m*9.81*sind(theta15); Rg30 =
   m*9.81*sind(theta30); %[N] grade Resistance
Fr0 = Ra+Rrl+Rg0; Fr15 = Ra+Rrl+Rg15; Fr30 = Ra+Rrl+Rg30;

% electric motor
Felpeak = Pelpeak./V;
Felpeak(Felpeak>6500)=6500;
Felnom = Pelnom./V;
Felnom(Felnom>4408)=4408;
%% plotting the results
figure(1);
plot(Gearspeed’,Torque’,’linewidth’,2);grid on;
xlabel(‘speed [km/h]’);ylabel(‘Torque [Nm]’)
figure(2);grid on; hold on;xlabel(‘speed [km/h]’);ylabel(‘Tractive Force [N]’);
   xlim([0 210]);ylim([0,7000])
%plot resistance lines
h1 = plot(V*3.6,Fr0,’g’,’linewidth’,2);plot(V*3.6,Fr15,’g’,’linewidth’,2);plot(V*3.6,
   Fr30,’g’,’linewidth’,2);
% plot lotus elise lines
h2 = plot(Gearspeed’,Force’,’k’,’linewidth’,2);
% plot electro motor lines
h3 = plot(V*3.6,Felpeak,’:r’,’linewidth’,2);
h4 = plot(V*3.6,Felnom,’b’,’linewidth’,2);
%set text
text(2,6000,’1st gear’);text(2,3200,’2nd gear’);text(2,2100,’3rd
   gear’);text(7,1600,’4th gear’);text(11,1200,’5th gear’)
text(177,3150,’30% slope’);text(177,2050,’15% slope’);text(177,900,’0% slope’);
xlim([0,200])
set(gcf,’color’,’w’);
set(gcf,’Units’,’centimeters’,’Position’,[2 2 20 10]);
%h_l = own_legend([h2(1);h3;h4;h1],{‘Lotus Elise S2’;’Electric motor (90kW/
   peak)’;’Electric motor (60kW/nom)’;’Driving Resistance’},10,’cm’,1,[0 0]);
h_l = own_legend([h1],{‘Driving Resistance’},10,’cm’,1,[0 0]);
set(h_l,’Position’,get(h_l,’position’)+[.65 .8 0 0])

%% Calculating the speed for the different engines for 0% slope
t = linspace(0,30,30001); tdelta = t(2);t_shift = 1;
gear = zeros(size(t));
velpeak = zeros(size(t));aelpeak = zeros(size(t));
velnom = zeros(size(t));aelnom = zeros(size(t));
vICE = zeros(size(t));vICE(1) = 9.28/3.6;aICE = zeros(size(t));
p1 = polyfit(Gearspeed(1,:)/3.6,Force(1,:),7);
p2 = polyfit(Gearspeed(2,:)/3.6,Force(2,:),7);
```

```
p3 = polyfit(Gearspeed(3,:)/3.6,Force(3,:),7);
p4 = polyfit(Gearspeed(4,:)/3.6,Force(4,:),7);
p5 = polyfit(Gearspeed(5,:)/3.6,Force(5,:),7);

for i=1:length(t)-1
    % electrical motor peak
    Felpeak = Pelpeak./velpeak(i);Felpeak(Felpeak>6500)=6500;
    % resistance forces
    Ra = rho/2*Cd*Af*velpeak(i).^2;  %[N] aerodynamic resistance
    frl = 0.01*(1+velpeak(i)/147);  %[0.01-0.015]
    Rrl = frl*m*9.81*2;        %[N] rolling resistance
    Rg = m*9.81*sind(0); %[N] grade Resistance
    % resulting force
    Fres = Felpeak-Ra-Rrl-Rg;
    % resulting acceleration
    aelpeak(i) = Fres/m;
    velpeak(i+1) = velpeak(i)+aelpeak(i)*tdelta;

    % electrical motor peak
    Felnom = Pelnom./velnom(i);Felnom(Felnom>4408)=4408;
    % resistance forces
    Ra = rho/2*Cd*Af*velnom(i).^2;  %[N] aerodynamic resistance
    frl = 0.01*(1+velnom(i)/147);  %[0.01-0.015]
    Rrl = frl*m*9.81*2;        %[N] rolling resistance
    Rg = m*9.81*sind(0); %[N] grade Resistance
    % resulting force
    Fres = Felnom-Ra-Rrl-Rg;
    % resulting acceleration
    aelnom(i) = Fres/m;
    velnom(i+1) = velnom(i)+aelnom(i)*tdelta;
    % ICE
    if vICE(i)<60.32/3.6  && gear(max(i-1,1))<2   %gear 1
      gear(i) = 1;
      FICE = polyval(p1,vICE(i));
    elseif vICE(i)<101/3.6  && gear(i-1)<3  %gear 2
      gear(i) = 2;
      FICE = polyval(p2,vICE(i));
    elseif vICE(i)<131/3.6  && gear(i-1)<4  %gear 3
      gear(i) = 3;
      FICE = polyval(p3,vICE(i));
    elseif vICE(i)<161/3.6  && gear(i-1)<5  %gear 4
      gear(i) = 4;
      FICE = polyval(p4,vICE(i));
    else                  %gear 5
      gear(i) = 5;
```

```
        FICE = polyval(p5,vICE(i));
    end
    if i>t_shift/tdelta && gear(i)~=gear(i-t_shift/tdelta)
        FICE = 0;
    end
    % resistance forces
    Ra = rho/2*Cd*Af*vICE(i).^2;  %[N] aerodynamic resistance
    frl = 0.01*(1+vICE(i)/147);  %[0.01-0.015]
    Rrl = frl*m*9.81*2;       %[N] rolling resistance
    Rg = m*9.81*sind(0); %[N] grade Resistance
    % resulting force
    Fres = FICE-Ra-Rrl-Rg;
    % resulting acceleration
    aICE(i) = Fres/m;
    vICE(i+1) = vICE(i)+aICE(i)*tdelta;
end
aICE(vICE>200/3.6)=0;
vICE(vICE>200/3.6)=200/3.6;

figure(3)
h=plot(t,velpeak*3.6,'r',t,velnom*3.6,'b',t+0.358,vICE*3.6,'k','linewidth',2);grid
  on;xlim([0 30]);
xlabel('time (s)');ylabel('velocity (km/h)')
set(gcf,'color','w');
set(gcf,'Units','centimeters','Position',[2 2 20 10]);
h_l = own_legend([h],{'electric motor 90kW';'electric motor 50kW';'Lotus
  Elise'},10,'cm',1,[0 0]);
set(h_l,'Position',get(h_l,'position')+[.65 .2 0 0])

figure(4)
h=plot(t,aelpeak,'r',t,aelnom,'b',t+0.358,aICE,'k','linewidth',2);grid on;xlim([0
  30]);
xlabel('time (s)');ylabel('acceleration (m/s^2)')
set(gcf,'color','w');
set(gcf,'Units','centimeters','Position',[2 2 20 10]);
h_l = own_legend([h],{'electric motor 90kW';'electric motor 50kW';'Lotus
  Elise'},10,'cm',1,[0 0]);
set(h_l,'Position',get(h_l,'position')+[.65 .75 0 0])
```

In most mobility applications, the designer will be provided (or need to derive) a duty-cycle (in mobility/vehicle terms, this is usually called a vehicle driving cycle). For the electric motor this is called a duty-cycle as this duty needs to be provided as a cycle to its shaft. Numerous vehicle-cycles already exist, e.g., FTP-75, NEDC, UDDS, HWFET, SAE J227a, EUDC,

ECE-15, JPN 10-15, CADC, etc., which then need to be "translated" to an electrical machine and drive duty-cycle. An example of a random duty-cycle shown in Figure 1.16.

The electrical machines and drives duty-cycle data needs to be formatted differently, where every interval in the duty-cycle must be described by: time duration, torque set point in [%] relative to rated (100Nm), start speed in [%] relative to rated (14150rpm), and stop speed in [%] relative to rated (14150rpm). In this manner the evaluated average torque was close to 30% (30Nm), and the evaluated average speed was close to 70% (9900rpm). The following were derived for this motor: peak torque = 101Nm, rated torque = 30Nm, maximum speed = 17700rpm, rated speed = 10200rpm, efficiency > 96%. An electrical machine that can fulfil this requirement is shown in Figure 1.17.

Using simulation software, the detailed magnetic and thermal analysis has to be undertaken, where the next figure provides an initial consideration as shown in Figure 1.18.

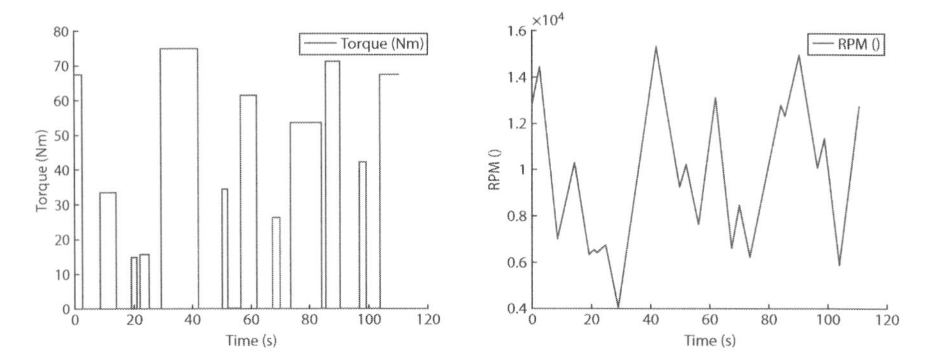

Figure 1.16 Example of a random duty-cycle is given by the following for a 2-wheeler application.

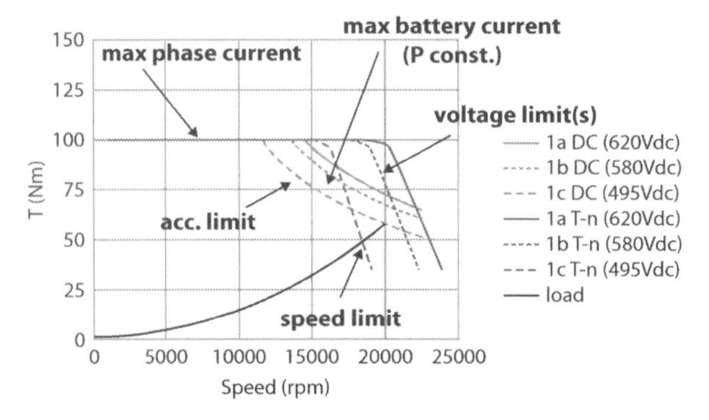

Figure 1.17 Example torque-speed characteristic for an electric 2-wheeler with a certain gear ratio [13].

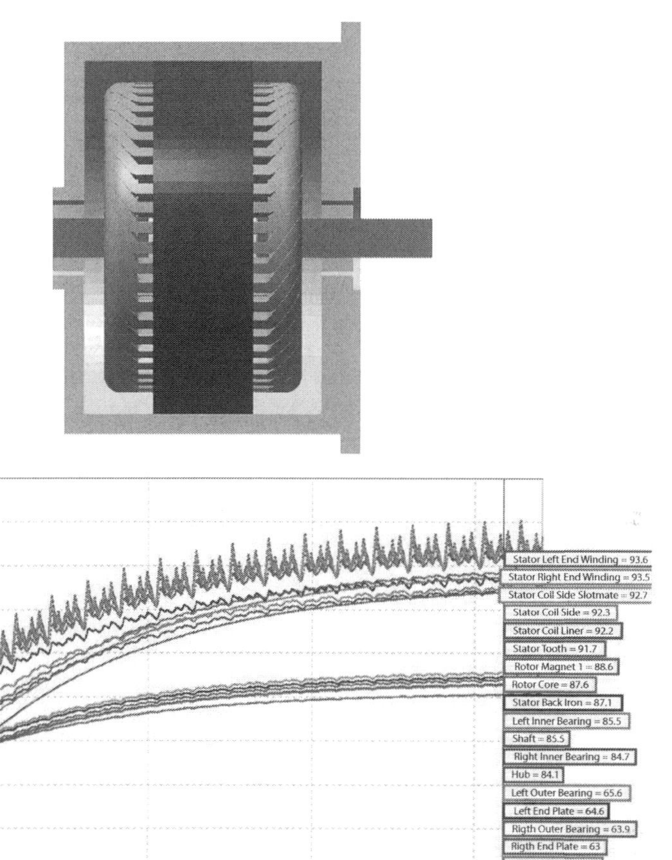

Figure 1.18 Finite Element Analysis and thermal results from the duty-cycle with coolest water temperature considered (courtesy Advanced Electromagnetics Group www.ae-grp.nl from the Netherlands).

This section has clearly identified that a constant power region can also be called field weakening, but is the base speed constant? What kind of field weakening exists? What is the definition?

1.7 Methods of Field Weakening Without a Clear Definition

What is field weakening and why is this necessary? Actually, numerous ways exist to achieve field weakening, which is a mere artefact to reduce the electrical machine and drive size for a vehicle. In the next figure a torque-speed characteristic of

1. Increasing the DC bus voltage
   - By increasing the DC bus voltage it is possible to achieve higher speeds, since the voltage limit is now moved (see figure). If voltage is electronically elevated (as in the Toyota Prius drive system), then the motor will run faster at an increased voltage for highway driving.
2. Battery voltage boosting using a DC-DC converter.
   - For example, the Prius system can boost the 200V battery voltage up to 500V for higher-speed operation.
3. Space Vector Modulation and its variants
   - This switching technique makes 15.5% more use of the available bus voltage thus also here the voltage limit is shifted.
4. Overmodulation (special form of space vector modulation)
   - This technique increases the amount of bus voltage; however, it does introduce significant harmonics in the voltage waveform.
5. Mechanical field weakening
   - The capability of high-speed motor controllers provides electronic-field weakening (bucking) as a control option; however; mechanical field weakening can also be used to overcome the limitations by sliding the magnet rotor axially along a splined motor shaft [14].
6. Commutation angle advance of the phase current with respect to the phase back-emf
   - Most of the existing literatures deem that interior permanent magnet machines are more appropriate for traction applications than surface mounted machines [15], because the interior permanent magnet motor rotor has more robust structure, additional reluctance torque and higher equivalent d-axis inductance. However, this argument is not persuasive. The surface mounted machines rotor can be designed with a high-strength retaining sleeve to overcome the mechanical issues and the stator slot area can be increased to allow a higher number of turns and hence higher d-axis inductance [16]. In addition, although surface mounted machines do not have saliency torque production, they have higher PM utilisation in the constant torque region and lower torque ripple over the whole speed range.

1.8 Consideration and Literature Concerning "Electronic" Field Weakening: What Does it Mean?

Even today, the separately excited DC commutator motor drive shows an ideal field-weakening characteristic. At low speed, rated dc motor current and rated flux are used to give rated performance. This voltage and output power both rise ideally linearly with speed. At rated speed, the voltage equals the rated dc terminal voltage. Above rated speed, the voltage is kept constant by decreasing the flux to the field winding (shunt coils), and hence the torque will be inverse proportion to speed. As the output power is constant this is also called the constant-power, field-weakening or flux-weakening region. There is no motor which has a flat output power characteristic against speed above the rated speed because of the iron and magnet losses. The separately excited DC machine has separate windings for the flux-producing and torque-producing currents. Brushless synchronous AC machines have a single stator winding which generates a rotating current phasor. For simplicity and with all assumptions mentioned before, this current phasor can be split into two components, i.e., the q-axis and d-axis current component. In permanent magnet machines the flux is mainly produced by the permanent magnets and flux-weakening is achieved by applying a negative current on the direct axis (d-axis). Every permanent magnet motor design has a particular, or most optimum, field-weakening control strategy to obtain maximum torque (and hence power) at any speed within the inverter volt-ampere rating [17]. This depends on the magnetic circuit, saturation level, voltage and current harmonics, etc. One of the primary limiting features of permanent magnet synchronous motor drives is the limited excitation control. The internal electromotive force (EMF) of the motor rises in proportion to the motor speed. Such behaviour is desirable in the so-called constant torque operating range. However, when the speed continues to rise, the voltage limit of the associated variable speed drive (VFD) is reached. The motor is then said to enter the flux-weakening operation, although already in the constant torque range also non-aligned q-axis current can be advantageous. Since the back EMF grows with the speed, the speed range can only be extended by reducing the air-gap flux, the so-called flux-weakening operation. Thus, the torque decreases while the speed increases, leading to keep the power constant – the so-called constant-power operation. In order to realize the flux-weakening control of the permanent magnet synchronous motor, the current and voltage vectors should be controlled in such a way that the d-axis armature current is negative, while the q-axis armature current is positive. It can be observed that the total flux linkage and hence the back EMF can be compensated by the induced voltage with negative d-axis armature current. Thus, by increasing i_d in the negative direction, the back EMF can be significantly weakened [18].

In addition to the constant torque requirement at low speeds, traction applications also require a powertrain that provides constant power at high speeds. At base speed the terminal voltage saturates to its maximum value. The speed of the motor cannot be increased any further unless a proper control strategy is implemented to reduce the stator flux. If a demagnetizing magnetomotive force (MMF) generated by manipulation of stator currents is applied, the apparent MMF generated by the permanent magnets could be reduced and the motor speed will increase. The strategy is known as field-weakening. Beyond base speed, the field weakening control takes over resulting in a simultaneous decrease in torque and an increase in speed. During the FW operation region, a demagnetizing MMF is established by the stator currents and winding to counteract the "apparent" MMF established by the permanent magnets (PMs) mounted on the rotor. As a result, the resultant air-gap flux is indirectly reduced/weakened and correspondingly the motor speed is increased [19]. It should be noted here that unlike in the maximum torque per Ampere control where the torque is only subjected to the current limit constraints, both the voltage and current constraints limit the torque production during the FW region [20].

Saliency (in stator or rotor) is required to produce a reluctance torque component, but to be apparent also requires the phase current to be advanced relative to the phase back-EMF. Phase advance depletes the so-called "magnet-alignment" torque, so there is a trade-off and an optimum phase-advance angle to maximize the total torque. The optimum angle is not fixed, but depends on the phase current, since saturation of the lamination (stator and rotor iron) causes a non-linear response. Therefore, the shaft torque in an optimized machine is rarely linearly proportional to the phase current [21]. For any magnitude of stator phase current, there is an optimum current excitation angle which leads to maximum electromagnetic torque. As speed increases, this angle is chosen for maximum torque per Ampere operation until the maximum phase voltage constrained by the DC-link is reached. Then the machine enters to field-weakening region where the current excitation angle is increased from the optimum value to 90 degrees electrical. Therefore, the d-axis component of the stator current vector is increased to reduce the air-gap flux [22].

Machines that are intended to operate over a very wide speed range present particular problems in relation to their inductance. Low inductance ensures that most of the available drive voltage can be expended in overcoming the EMF, but motoring speed cannot then exceed the speed at which the EMF becomes equal to the available drive voltage. If the machine has sufficient inductance, the armature-reaction flux can be used to suppress the magnet flux to some degree, reducing the induced voltage and permitting the motor to run faster. The phase shift in the current required to achieve this flux-weakening causes a reduction in the torque constant, so this mode of operation is often associated with the search for a constant-power operating characteristic, rather than constant torque (versus speed) [23].

1.9 Summary of Electronic Field Weakening Definitions

In literature sources, the definition of field weakening control is likely to be concentrated around one of the following descriptions:

1. The operating speed range can be extended by applying negative field stator current component to weaken the airgap flux [24, 25].
2. Flux weakening control increases a motor's speed range by reducing the flux density and flux in its air-gap through the application of a demagnetizing armature current component along the d-axis of the permanent magnets [21, 26, 27].
3. To improve the speed of the machine above base speed, while maintaining the motor terminal voltage constant, this can be done by regulating id, iq, which is so-called flux weakening control [28].
4. When the voltage limit of the associated frequency converter is reached, the motor is then said to enter the flux weakening operation. The internal voltage must now be adjusted to be compatible with the applied converter voltage which increases as speed increases. As a result, the motor power factor becomes leading and the current to be commutated by the inverter continues to increase as speed increases [29].
5. Armature-reaction flux can be used to suppress the magnet flux to some degree, reducing the induced voltage and permitting the motor to run faster [23].
6. It can be observed that the total flux linkage and hence the back EMF can be compensated by the induced voltage with negative d-axis armature current. Thus, by increasing id in the negative direction, the back EMF can be significantly weakened [18].
7. Field weakening control is a strategy to obtain maximum torque (and hence power) at above base speed within the inverter volt-ampere rating [17].

8. Optimal design of the flux weakening strategy aims at guaranteeing the maximum available voltage exploitation, thus providing the maximum torque for any given speed and minimizing the phase current magnitude (therefore reducing one of the main power losses) [30].
9. For a feasible torque below rated machine torque and angular velocities higher than a certain feasible MTPA velocity, the machine is operated in FW. In order to improve this and extend the speed range, it is necessary to maximize the IPMSM torque output appropriately along the optimal current trajectory over the whole speed range [31, 32].
10. Phase advance, advancing the phase current in respect to the back EMF, weakens the d-axis flux, making higher speeds possible with a given drive voltage. This is the essence of flux weakening [12].
11. Demagnetizing MMF is established by the stator currents and winding to counteract the "apparent" MMF established by PMs. Resultant air-gap flux is indirectly reduced/weakened and correspondingly the motor speed is increased [19, 20].
12. By increasing the excitation current angle, the machine operates in field weakening region and the relative electrical position of armature pole and PM pole is shifted by the amount of excitation current angle. Hence most of the PM surface is faced to the armature pole. Since the armature pole and PM pole have opposite directions, the resultant field is weakened [16, 22].

1.10 Critical Study of Field Weakening Definitions

- "Field weakening" or "flux weakening"

These terms can be used interchangeably, as they represent the same thing.

- *"The operating speed range can be extended" or "permitting the motor to run faster" or "Flux weakening control increases a motor's speed range" or "allowing for velocities higher than MTPA velocity"*

The terms described above do not define the field weakening region in great detail, i.e., "field weakening control enables the machine to operate above the base speed". At base speed, the terminal voltage equals the rated terminal voltage that is determined by the fixed DC-link voltage. Above base speed, the terminal voltage is kept constant by decreasing the flux, and hence the torque, in inverse proportion to speed. As the output power is constant this is called the constant-power, field weakening or flux weakening region.

Real electrical machines do not have constant output power above base speed mainly due to iron losses; therefore the constant-power speed range (CPSR) is defined as the speed range over which rated power can be maintained.

- "*regulating id, iq, is so-called field weakening control, through increasing the direct-axis demagnetization current*" or "*applying negative field stator current*" *or* "*by applying a demagnetizing armature current component along the d-axis of the permanent magnets*".

These are vague descriptions of what is happening during field weakening. In theory, by using the dq-reference frame transformation for flux linkages

$$\Psi_s = \sqrt{\Psi_d^2 + \Psi_d^2} \tag{1.1}$$
$$\Psi_d = \Psi_m + L_d i_d$$

with term $L_d i_d$ being the armature current component along the d-axis, the term $L_d i_d$ is demagnetizing when the applied i_d current is negative, where the d- and q-axis are defined as shown in Figure 1.19. This in turn weakens the d-axis flux linkage thereby decreasing the flux linkage vector, which in turn reduces the back electromotive force induced in the coils.

However, the dq-axis reference frame, also often referred to as the pseudo dq-axis reference frame due to the sole consideration of the fundamental harmonic of the phase voltages and currents, is not an accurate frame to describe the essence of field weakening. This is because, in terms of the machine field weakening capability, voltage distortion and ripple are rarely considered and explained, as shown in Figure 1.20. For traction applications, sinusoidal phase current is desired to enhance machine torque performance and variable speed drives employing vector control method are commonly used for traction machine supply. The basic operating theory of classical synchronous machines is the simplified phasor diagram based on the pseudo

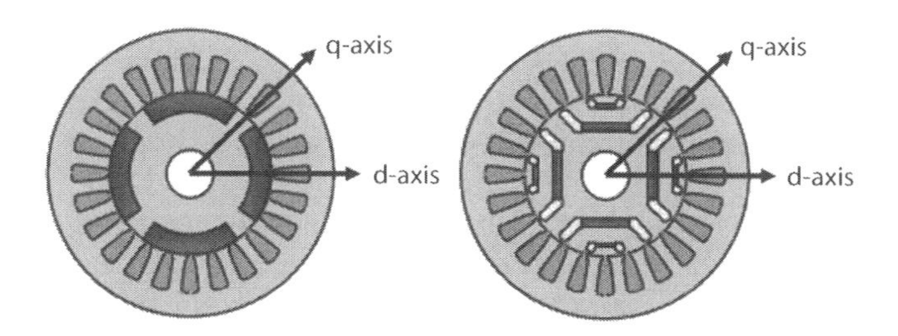

Figure 1.19 Illustrating the d- and q-axis in a permanent magnet machine with a varying level of saliency in the present in the rotor lamination (iron core).

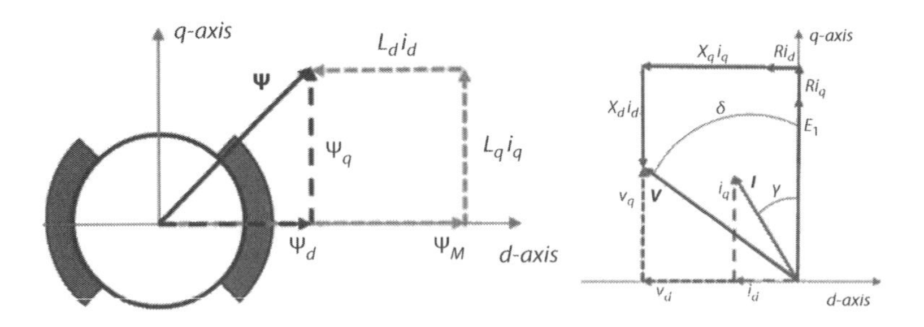

Figure 1.20 Field weakening and phasor diagram.

dq-axis reference frame, and digital variable speed drives are usually based on vector (field oriented) principles, although there are many varieties. Most of them nowadays are based on the assumption of sinusoidal electromotive force, and considerable effort is expended to achieve this in practice. The current is mostly sinusoidal at low and medium speeds, but at high speeds harmonics appear as the current regulator saturates or "runs out of voltage". All permanent magnet machines suffer from phase voltage and back EMF distortion with increasing speed, especially in the field weakening region, due to magnetic saturation, machine geometry and armature reaction caused by the interaction of the permanent magnet flux and armature flux distributions, hence suffer from magnetic nonlinearities and variations in the leakage flux.

In essence, during field weakening, the fundamental of the phase current, represented by i_d and i_q via the Park and Clarke transformation, is advanced in phase with respect to the back electromotive force, E_I. The advancing of the phase current establishes a demagnetizing armature MMF, $X_d i_d$, to counteract the MMF of the permanent magnet flux linkage, Ψ_M, which in turn lowers the net flux flowing. In other words, by increasing the commutation angle, defined as the angle between the fundamental of the phase current and the fundamental of the phase back electromotive force, the machine operates in field weakening region and the relative electrical position of armature pole and permanent magnet pole is shifted by the amount of commutation angle. Hence the permanent magnet surface is faced partly to the armature pole. Since the armature pole and PM pole have opposite directions, the resultant field is weakened.

- "*weaken the airgap flux*" or "*by reducing the flux density and flux in its air-gap*"

The airgap flux itself is not weakened, but this is a result of the reduction in the flux flowing in the magnetic circuit which is a result of the demagnetizing armature MMF that counteracts the MMF of the permanent magnets. Due to the lowered flux, the phase back EMF induced in the coils is also reduced:

$$e_{ph} = \frac{d\Psi_M}{dt} \tag{1.2}$$
$$\Psi_M \approx N\Phi.$$

Since $\Phi_m = \Phi_g + \Phi_L$ and if Φ_m is lowered, the airgap flux also reduces, which means that the flux density in the airgap is reduced thereby as well.
• "*improving the speed of the machine above base speed, while maintaining the motor terminal voltage constant, can be done by regulating id, iq*" or "*field weakening control is a strategy to obtain maximum torque (and hence power) at above base speed within the inverter volt-ampere rating*" or "*Optimal design of the flux weakening strategy aims at guaranteeing the maximum available voltage exploitation, thus providing the maximum torque for any given speed and minimizing the phase current magnitude (therefore reducing one of the main power losses).*" Or "*it is necessary to maximize the combination of magnet alignment and reluctance torque appropriately along the optimal current trajectory over the whole speed range*".

Above base speed, feasible operating region of the machine lies within the intersection of the maximum voltage ellipse and the maximum current "circle" (the question is really whether this is a circle in any practical permanent magnet electric drive application). The maximum voltage ellipse is inversely proportional with speed such that the feasible operating area shrinks. The strategy in this feasible operating area is done by choosing *wisely* the amount of d- and q-current, or in other words, choosing *wisely* how much the phase current will be advanced with respect to the phase back EMF. Furthermore, strategies in field weakening region depend on speed and torque demands. There are several methods of field weakening control, of which feedforward and feedback form the base. These algorithms are often designed by the intersection of the demanded torque and voltage limit (FW strategy), or if the demanded torque cannot be reached anymore, the intersection of the voltage limit and current limit. There are also objectives to obtain maximum torque per voltage (MTPV). Thus there are more strategies in the field weakening region to comply to a certain operation point in the best way.

One of the primary limiting features of permanent magnet synchronous motor drives is the limited excitation control. The internal electromotive force (EMF) of the motor rises in proportion to the motor speed as defined in 2-7. Such behaviour is desirable in the so-called constant torque operating range. However, when the speed continues to rise, the voltage limit of the associated variable speed drive is reached. The motor is then said to enter the field weakening operation. It should be noted here that unlike in the constant torque region where the torque is only subjected to the current limit constraints, both the voltage and current constraints limit the torque production during the FW region. At base speed the voltage phasor saturates to its maximum value (see Figure 1.20), due to the EMF. The speed of the motor cannot be increased any further unless a proper control strategy is implemented to reduce the stator flux and therefor compensating the induced voltage.

In order to realize the field weakening control of the permanent magnet synchronous motor, the current and voltage vectors should be controlled in such a way that the phase current waveform gets phase advanced with respect to the back EMF by the current/commutation angle γ (Figure 1.20)

$$\begin{aligned} \mathrm{i_d} &= -\boldsymbol{I}\sin\gamma; \\ \mathrm{i_d} &= \boldsymbol{I}\cos\gamma. \end{aligned} \tag{1.3}$$

This is done by introducing a negative d-axis armature current, since then the back EMF, as one might suspect from above equation gets compensated by the demagnetizing term as visualized in Figure 1.20. Thus, by increasing in the negative direction, the flux linkage term $L_d \cdot i_d$ visualized in Figure 2-4 opposes the permanent magnet flux linkage resulting in significant flux/field weakening, hence the name of this strategy is called flux or field weakening control. As a result of the weakening of flux, the torque decreases while the speed increases, leading to keep the power constant (see Figure 2-2), the so-called constant-power operation [17]. Machines that are intended to operate over a very wide speed range present particular challenges in relation to their inductance and maximum current. Low inductance ensures that most of the available drive voltage can be expended in overcoming the EMF, but motoring speed cannot then exceed the speed at which the EMF becomes equal to the available drive voltage. If the machine has sufficient inductance and current, the armature-reaction flux can be used to suppress the magnet flux sufficiently.

Every permanent motor design has a particular field weakening control strategy, or most optimum angle , to obtain maximum torque (and hence power) at any speed within the inverter volt-ampere rating. This depends on the magnetic flux circuit, saturation level, saliency, voltage and current harmonics [16] etc.

1.11 Motor Limits

Any electrical drive controller concerning optimal operating capability over a wide range of speed is limited by motor drive limits. A good guide for obtaining this capability is to focus on the current limit circle and the voltage limit ellipse.

The current limit circle is a circular boundary in the i_d, i_q plane which does not exceed the current limit I_{max}

$$\sqrt{i_d^2 + i_q^2} = I_{max} \tag{1.4}$$

which depending on the inverter size and machine construction, can represent either the maximum current available from the inverter or the maximum permittable current of the motor.

The voltage limit ellipse is a **current** locus in the i_d, i_q plane which defines all possible currents that can be obtained at the specified inverter voltage limit, rather than its current, by V_{max} (Hendershot & Miller, 2010)

$$\left(X_q \cdot i_q\right)^2 + \left(E_1 + X_d \cdot i_d\right)^2 = \left(V_{max}\right)^2 \tag{1.5}$$

which is the maximum fundamental phase voltage available from the inverter, and it depends on the particular PWM modulation index m (0.5 for PWM) with the DC bus voltage V_{dc}

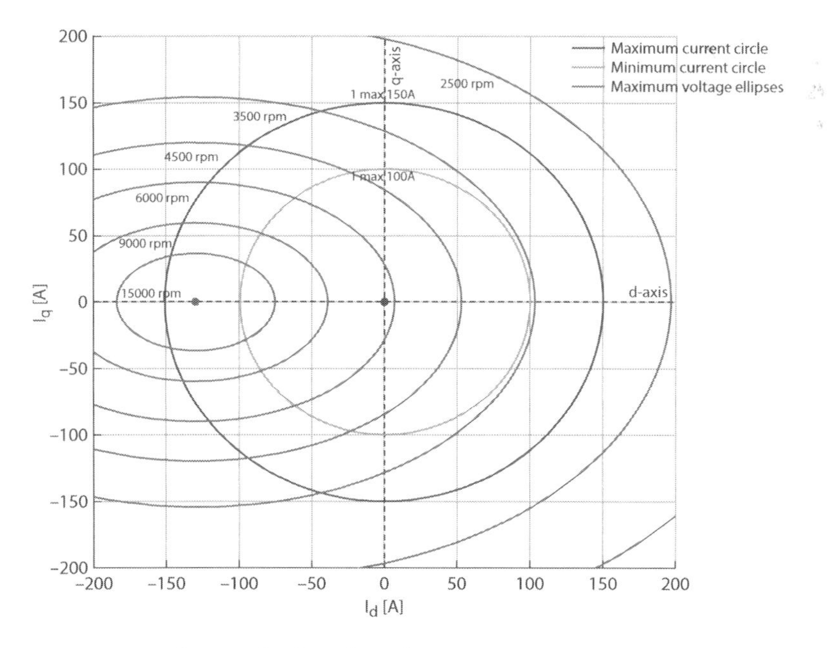

Figure 1.21 Current limit and voltage limit diagram.

$$V_{max} = m \cdot V_{dc} \qquad 1.7$$

If $X_d = X_q$, i.e., nonsalient machines, the voltage limit becomes a circle. The voltage limit shrinks with the increase of speed. Operating points of the electrical machine can be visualized in the i_d, i_q plane with the help of the current and voltage limit diagrams as shown in Figure 1.21.

```
clc;close all;clear
%% drive
I_max      = 150;      % [A] maximum current, this is used in the velocity
   controller
V_dc       = 40;       % [V] DC bus voltage for reference to create the PWM
   reference
PWM_mod      = 0.5;      % [-] PWM modulation index
V_max      = V_dc*PWM_mod; % [V] maximum voltage
p          = 4;        % [-] pole pairs number
Rs         = 30e-3;      % [Ohm] stator phase resistance cold - from SPEED
Psi_PM      = 0.0076;     % [wb] rms PM flux linkage - from SPEED
Lq         = 8.8697*10^-5; % [H] q-axis inductance
Ld         = 5.8112*10^-5; % [H] d-axis inductance
m          = 3;        % [-] number of phases
%% current and voltage diagram (source is book hendershot and miller)
figure(1);grid on; hold on;xlabel('i_d [A]');ylabel('i_q [A]');
yl = yline(0,'--','d-axis','LineWidth',2);
xl = xline(0,'--','q-axis','LineWidth',2);
C_ellipse = plot(-Psi_PM/Ld,0,'r*','LineWidth',2); %center point of voltage
   ellipse
C_circle = plot(0,0,'b*','LineWidth',2); %center point of current circle
xlim([-I_max-50 I_max+50]);
ylim ([-I_max-50 I_max+50]);
```

```
% I_max limit 1
ang = [0:0.01:(2*pi)];
iq_Imax = I_max * sin(ang);
id_Imax = I_max * cos(ang);
plot(id_Imax,iq_Imax,'b','linewidth',1);

% I_max limit 1
ang = [0:0.01:(2*pi)];
iq_Imax2 = (I_max-50) * sin(ang);
id_Imax2 = (I_max-50) * cos(ang);
plot(id_Imax2,iq_Imax2,'c','linewidth',1);

% V_max limit ellipse equation book miller page 334 and 338
%(Lq*we*Iq)^2+(w*Psi_PM+(Ld*we*Id))^2=V_max^2
rpm = [2000 2500 3500 4500 6000 9000 15000];
we = rpm*p*pi/30;
for ii = 1:length(we)
ang2 = [0:0.01:(2*pi)];
we_s = we(ii);
```

```
Xd = Ld*we_s;
Xq = Lq*we_s;
E1 = Psi_PM*we_s;
id_Vmax = (V_max*cos(ang2)-E1)/Xd;
iq_Vmax = V_max*sin(ang2)/Xq;
plot(id_Vmax,iq_Vmax,'r','linewidth',1);hold on;
end

%set text
text(160,165,'2000 rpm');text(40,165,'2500 rpm');text(-100,140,'3500
   rpm');text(-150,110,'4500 rpm');text(-165,80,'6000 rpm');text(-178,45,'9000
   rpm');text(-183,5,'15000 rpm');
text(-20,155,'I max 150A');text(-20,95,'I max 100A');
% xlim([0,200])
set(gcf,'color','w');
set(gcf,'Units','centimeters','Position',[2 2 24 20]);

%% current and voltage diagram with Saliency optmisation (source is book
   hendershot and miller)
figure(2);grid on; hold on;xlabel('i_d [A]');ylabel('i_q [A]');
% yl = yline(0,'--','d-axis','LineWidth',2);
% xl = xline(0,'--','q-axis','LineWidth',2);
C_ellipse = plot(-Psi_PM/Ld,0,'r*','LineWidth',2); %center point of voltage
   ellipse
C_circle = plot(0,0,'b*','LineWidth',2); %center point of current circle
xlim([-I_max-50 I_max+50]);
ylim ([-I_max-50 I_max+50]);

% I_max limit 1
ang = [0:0.01:(2*pi)];
iq_Imax = I_max * sin(ang);
id_Imax = I_max * cos(ang);
plot(id_Imax,iq_Imax,'b','linewidth',1);

% I_max limit 1
ang = [0:0.01:(2*pi)];
iq_Imax2 = (I_max-50) * sin(ang);
id_Imax2 = (I_max-50) * cos(ang);
plot(id_Imax2,iq_Imax2,'c','linewidth',1);

% V_max limit ellipse equation book miller page 334 and 338
%(Lq*we*Iq)^2+(w*Psi_PM+(Ld*we*Id))^2=V_max^2
rpm = [2000 2500 3500 4500 6000 9000 15000];
we = rpm*p*pi/30;
for ii = 1:length(we)
ang2 = [0:0.01:(2*pi)];
```

```
we_s = we(ii);
Xd = Ld*we_s;
Xq = Lq*we_s;
E1 = Psi_PM*we_s;
id_Vmax = (V_max*cos(ang2)-E1)/Xd;
iq_Vmax = V_max*sin(ang2)/Xq;
plot(id_Vmax,iq_Vmax,'r','linewidth',1);hold on;
end

% MTPA Salient
% for positive q currents
we = 1 % angular velocity, this is not dependant for the MTPA since it stripes
   away with the div
E1 = Psi_PM*we;
Xd = Ld*we;
Xq = Lq*we;
I = [0:1:I_max]
dV = (Xd-Xq).*I;
gamma_Tmax = asind(1/4*(E1./dV+sqrt((E1./dV).^2+8))); % torque angle to
   get maximum torque per amp
id_MTPA = -I.*sind(gamma_Tmax);
iq_MTPA = I.*cosd(gamma_Tmax);
plot(id_MTPA,iq_MTPA,'g','linewidth',1);

% MTPA Salient
% for negative q currents
we = 1 % angular velocity, this is not dependant for the MTPA since it stripes
   away with the div
E1 = Psi_PM*we;
Xd = Ld*we;
Xq = Lq*we;
I = [0:1:I_max];
dV = (Xd-Xq).*I;
gamma_Tmax = asind(1/4*(E1./dV+sqrt((E1./dV).^2+8))); % torque angle to
   get maximum torque per amp
id_MTPA = -I.*sind(gamma_Tmax);
iq_MTPA = -1.*I.*cosd(gamma_Tmax);
plot(id_MTPA,iq_MTPA,'g','linewidth',1);

% % MTPA non Salient
I = [-I_max:1:I_max];
gamma_Tmax_ns = 0; % torque angle to get maximum torque per amp
id_MTPA = -I.*sind(gamma_Tmax_ns);
iq_MTPA = I.*cosd(gamma_Tmax_ns);
% iq_MTPA = [-I_max+10:1:I_max-10];
```

```
% id_MTPA = (-Psi_PM + sqrt(Psi_PM^2+4*(Ld-Lq)^2.*iq_MTPA.^2))/
   (2*(Ld-Lq));
plot(id_MTPA,iq_MTPA,'m','linewidth',1);

%set text
text(160,165,'2000 rpm');text(40,165,'2500 rpm');text(-100,140,'3500
   rpm');text(-150,110,'4500 rpm');text(-165,80,'6000 rpm');text(-178,45,'9000
   rpm');text(-183,5,'15000 rpm')
ext(-20,155,'I max 150A');text(-20,95,'I max 100A');
% xlim([0,200])
set(gcf,'color','w');
set(gcf,'Units','centimeters','Position',[2 2 24 20]);

%% current and voltage diagram with Saliency optmisation (source is book
   hendershot and miller)
figure(3);
subplot(211);grid on; hold on;xlabel('i_d [A]');ylabel('i_q [A]');
% yl = yline(0,'--','d-axis','LineWidth',2);
% xl = xline(0,'--','q-axis','LineWidth',2);
t% C_ellipse = plot(-Psi_PM/Ld,0,'r*','LineWidth',2); %center point of voltage
   ellipse
C_circle = plot(0,0,'b*','LineWidth',2); %center point of current circle
xlim([-I_max-50 0]);
ylim ([0 I_max+50]);
% I_max limit 1
ang = [0:0.01:(2*pi)];
iq_Imax = I_max * sin(ang);
id_Imax = I_max * cos(ang);
plot(id_Imax,iq_Imax,'b','linewidth',1);
% I_max limit 1
ang = [0:0.01:(2*pi)];
iq_Imax2 = (I_max-50) * sin(ang);
id_Imax2 = (I_max-50) * cos(ang);
plot(id_Imax2,iq_Imax2,'c','linewidth',1);
% MTPA Salient
% for positive q currents
we = 1 % angular velocity, this is not dependant for the MTPA since it stripes
   away with the div
E1 = Psi_PM*we;
Xd = Ld*we;
Xq = Lq*we;
I = [0:1:I_max];
dV = (Xd-Xq).*I;
gamma_Tmax = asind(1/4*(E1./dV+sqrt((E1./dV).^2+8))); % torque angle to
   get maximum torque per amp
```

```
id_MTPA = -I.*sind(gamma_Tmax);
iq_MTPA = I.*cosd(gamma_Tmax);
plot(id_MTPA,iq_MTPA,'g','linewidth',2);
% % MTPA non Salient
I = [-I_max:1:I_max];
gamma_Tmax_ns = 0; % torque angle to get maximum torque per amp
id_MTPA = -I.*sind(gamma_Tmax_ns);
iq_MTPA = I.*cosd(gamma_Tmax_ns);
% iq_MTPA = [-I_max+10:1:I_max-10];
% id_MTPA = (-Psi_PM + sqrt(Psi_PM^2+4*(Ld-Lq)^2.*iq_MTPA.^2))/
   (2*(Ld-Lq));
plot(id_MTPA,iq_MTPA,'m','linewidth',2);
% Torque loci
%T = (m*p/we).*(id + ((Psi_PM*we)/(Ld*we - Lq*we))).* iq *(Ld*we - Lq*we)
T = [0:1:8];
for ii = 1:length(T)
T_s = T(ii);
id_T = [-I_max-50:1:I_max+50];
iq_T = T_s./(3/2*p*(Psi_PM+id_T*(Ld-Lq)));
plot(id_T,iq_T,'k','linewidth',1);hold on
end
%set text
text(-180,110,'T = 8 Nm');text(-180,97,'T = 7 Nm');text(-180,84,'T = 6
   Nm');text(-180,70,'T = 5 Nm');text(-180,57,'T = 4 Nm');text(-180,45,'T = 3
   Nm');text(-180,31,'T = 2 Nm');text(-180,18,'T = 1 Nm');text(-180,5,'T = 0
   Nm');
text(-20,155,'I max 150A');text(-20,95,'I max 100A');
% xlim([0,200])
set(gcf,'color','w');
set(gcf,'Units','centimeters','Position',[2 2 24 20]);

% torque vs torque angle
deg = char(176);
subplot(212);grid on; hold on;ylabel('electromagnetic torque T_e
   [Nm]');xlabel('commutation angle \gamma [{\circ}]' );
hold on;grid on
we = 500 % angular velocity, this is not dependant for the MTPA since it stripes
   away with the div
E1 = Psi_PM*we;
Xd = Ld*we;
Xq = Lq*we;
I = [0:1:I_max];
```

```
dV = (Xd-Xq).*I;
gamma_Tmax = asind(1/4*(E1./dV+sqrt((E1./dV).^2+8))); % torque angle to
    get maximum torque per amp
% optimum torque angle
for ii = 1:length(gamma_Tmax)
gamma_Tmax_s = gamma_Tmax(ii);
I = [0:1:I_max];
Te = 3/2*p*(Psi_PM.*I*cosd(gamma_Tmax_s)-I.^2*sind(gamma_
    Tmax_s)*cosd(gamma_Tmax_s)*(Ld-Lq));
end
plot(gamma_Tmax,Te,'g','linewidth',2);
% id=0
gamma_Tmax_id_0 = zeros(1,151)
for ii = 1:length(gamma_Tmax_id_0)
gamma_Tmax_s = gamma_Tmax_id_0(ii);
I = [0:1:I_max];
Te = 3/2*p*(Psi_PM.*I*cosd(gamma_Tmax_s)-I.^2*sind(gamma_
    Tmax_s)*cosd(gamma_Tmax_s)*(Ld-Lq));
end
plot(gamma_Tmax_id_0,Te,'m','linewidth',2);
% torque vs torque angle
I_var = [0:25:I_max];
for ii = 1:length(I_var)
    hold on
gamma = [-90:1:90];
I_var_s = I_var(ii);
Te_var = 3/2*p*(Psi_PM.*I_var_s.*cosd(gamma)-I_
    var_s.^2.*sind(gamma).*cosd(gamma).*(Ld-Lq));
plot(gamma,Te_var,'k','linewidth',1); hold on
end
hold on
% rel and mag torque for max current over angle
gamma = [-90:1:90];
Te_mag = 3/2*p*(Psi_PM.*I_max.*cosd(gamma))
Te_rel = 3/2*p*(-1*I_max^2.*sind(gamma).*cosd(gamma).*(Ld-Lq))
plot(gamma,Te_mag,'b','linewidth',2); hold on
plot(gamma,Te_rel,'r','linewidth',2); hold on
%set text
text(35,8,'I = 150 A');text(33,6.5,'I = 125 A');text(21,5.2,'I = 100 A');text(19,3.9,'I
    = 75 A');text(15,2.7,'I = 50 A');text(12,1.7,'I = 25 A');text(12,0.5,'I = 0 A');
set(gcf,'color','w');
set(gcf,'Units','centimeters','Position',[2 2 18 18]);
```

1.12 Concluding Remarks

In the pursuit of reducing global energy demand, maximizing the efficiency of electric machines and drives is of paramount importance, as more than half or even two-thirds of today's generated electricity is converted to mechanical by electric machinery. Today, electrical mobility is something we take for granted on our roads, since it can minimize local exhaust pollution to an energy generation location. Only if electricity becomes a truly sustainable source will we all benefit. Electrical energy is advantageous since it can be "in principle" transported with almost 100% efficiency over relatively long, interconnected distances. The challenge is about energy storage and mostly has to be converted to another energy form to allow bidirectional utilization. This remains one of the unsolved challenges of the 21st century!

Urbanization, green thinking, and self-driving vehicles are just a few of the contemporary trends that are changing the face of transportation and the overall mobility system. The automotive industry in particular needs to rethink its business models in order to continue to provide solutions that address current and future trends. A shift from being a traditional car manufacturer with a business model focused on product sales toward being a provider of mobility services seems to be one promising approach among many other possibilities. Furthermore, stricter legislative restrictions result in new challenges that are pushing vehicle development engineers to adopt highly efficient and effective solutions. The automotive powertrain is undergoing a transformation due to the increased electrification. This transformation brings a number of technical challenges relating to the design and integration of electrified elements into the powertrain. Powertrain-related targets can be achieved in an increasing number of ways by varying and balancing the powertrain elements. Powertrain engineers will face new challenges in the development and production of internal combustion engines, e-drives, transmissions, batteries, capacitors, tyres, body shapes, and fuel cells, and how a systems engineering approach is the key to understanding and taking advantage of all the possibilities in such complex electrified powertrains. The complexity of a vehicle and also of its subsystems such as powertrain, electrics/electronics (E/E), thermal management system, chassis, body, or driving assistance is increasing due to the growing number of interacting functions and (mechatronic) systems. This results in the need to continuously coordinate, monitor, adjust, and optimize development tasks across departments, company boundaries and even globally. Also, there is a need to consider the various regulations as have been described in this chapter, ranging from vehicle classes to machine directive. This Directive contains a list of essential health and safety requirements related to the design and construction of machinery to allow a CE mark to be used on present and future vehicles.

In particular for permanent magnet synchronous machines with symmetrical or non-symmetrical rotor designs, the challenge is to provide an optimum efficiency both in the constant torque as in the constant power region of the torque-speed characteristic envelope of the powertrain. In this respect, every electrical machine and drive specialist should try to exploit the ambiguity in the selection of the stator current's direct and quadrature components producing the highest amount of torque or efficiency, such that losses are minimized while physical constraints are satisfied (e.g., speed, vibration, torque ripple, current or voltage limits). Depending on the actual operating conditions, different optimization problems may be formulated leading to respective operation strategies, such as maximum torque per current (MTPC, also known as MTPA), maximum torque per voltage (MTPV), maximum current (MC), maximum torque per losses (MTPL) and/or field weakening (FW).

As described above, definitions of field weakening vary in literature. Field weakening is first of all more complex than "just" regulating a negative d-axis current. Due to harmonics the pseudo dq-axis reference frame is not sufficient to define field weakening. The feasible field weakening region is limited by variable speed drive current and voltage limits. Furthermore, machine nonlinearities and losses set further requirements to keep in mind during the field weakening region. In these final chapter remarks we also attempt to provide a more general definition of field weakening:

> Field weakening is the strategy to find optimum commutation angles, with a given objective, covering motors running at all speeds. The control algorithm enables electrical machines to operate optimal in all running conditions by making use of saliency and therefore phase advancing the current with respect to the back EMF appropriately. This also includes, for example, increasing the torque for a given current below base speed, and optimal control in the field weakening region, which is defined as the speed above base speed. Or voltage minimization considering higher order harmonics into the phase current waveform accounting for the limitations in phase currents and voltages. Furthermore, machine nonlinearities, copper losses and iron losses set further requirements to keep in mind during the implementation of this algorithm.

References

1. https://pixabay.com/illustrations/background-pattern-brain-face-5127760/. Pictures taken from pixabay Stunning free images & royalty free stock
2. https://wallpaperaccess.com/the-flintstones Pictures taken from free images & royalty free stock
3. https://www.un.org/en/development/desa/population/publications/pdf/trends/Population2030.pdf
4. Photo courtesy of Brandweer Nederland, https://zerauto.nl/achtergrond-daarom-plaatst-de-brandweer-sommige-autos-in-een-waterbak/
5. Photo courtesy of Bike Europe, https://www.bike-eu.com/home/nieuws/2018/02/underrated-e-bike-battery-fire-hazards-call-for-attention-on-safe-storage-10132804
6. https://ec.europa.eu/clima/policies/transport_en
7. https://www.mckinsey.com/industries/automotive-and-assembly/our-insights/the-irresistible-momentum-behind-clean-electric-connected-mobility-four-key-trends
8. https://eur-lex.europa.eu/LexUriServ/LexUriServ.do?uri=OJ:L:2013:060:0052:0128:EN:PDF
9. https://eur-lex.europa.eu/legal-content/EN/TXT/PDF/?uri=CELEX:32019R0129&from=EN
10. https://pixabay.com/photos/bike-cycling-cyclists-movement-5045324/
11. https://pixabay.com/nl/vectors/lege-batterij-lek-lekkage-gratis-1623377/, https://pixabay.com/nl/illustrations/batterij-pictogrammen-set-van-iconen-3201720/
12. Hendershot, J., & Miller, T. (2010). *Design of brushless permanent-magnet machines.* Florida: Motor Design Books LLC.
13. https://electricsuperbiketwente.nl/
14. Zepp L.P. "Advantages of Mechanical vs. Electronic-Field Weakening." Dura-Trac Motors, Inc. www.duratracmotors.com
15. Lu D., Kar N. C., (2010) "A Review of Flux-weakening Control in Permanent Magnet Synchronous Machines", DOI:10.1109/VPPC.2010.5728986
16. Zhao, N., Schofield, N., & Hu, Y. (2019). "Phase voltage distortion of IPM and SPM machines with distributed windings in field weakening region", *Journal of Engineering*, Vol. 2019, Issue 17, DOI: 10.1049/joe.2018.8125, Online ISSN 2051-3305.
17. Soong, W., & Miller, T. (1994). "Field-weakening performance of brushless synchronous AC motor drives", *IEE Proceedings - Electric Power Applications*, Vol. 141, Issue 6, DOI:10.1049/IP-EPA:19941470Corpus ID: 108708275
18. K.T. Chau. (2015). *Electric vehicle machines and drives*, Hong Kong: John Wiley & Sons Singapore Pte. Ltd. (p. 88)
19. A.H. Ünsal. (2016). "Maximum Torque per Ampere and Flux Weakening control of IPM motors for electrical vehicles". Istanbul: Istanbul Technical University, https://polen.itu.edu.tr/xmlui/handle/11527/15614

20. Khan, W. (2016). "Torque Maximizing and Flux Weakening Control of Synchronous Machines". Aalto University, https://aaltodoc.aalto.fi/handle/123456789/20877
21. Aalborg University. (2009). "Torque Control in Field Weakening Mode". Master thesis, Aalborg University, Institute of Energy Technology, https://projekter.aau.dk/projekter/en/studentthesis/torque-control-in-field-weakening-mode(c3e7c0a4-bc9b-4542-9b01-15f6c37b745e).html (page 42)
22. Zhao, N. (2017). "Modeling and design of electric machines and associated components for more eleectric vehicles". Ontario: McMaster University, https://macsphere.mcmaster.ca/handle/11375/21466
23. Hendershot, J., & Miller, T. (2010). "Design of brushless permanent-magnet machines, Florida: Motor Design Books LLC. (p. 213).
24. Chi, S., & Xu, L. (2006). "A Special Flux-weakening Control Scheme of PMSM - Incorporating and Adaptive to Wide-Range Speed Regulation", 2006 CES/IEEE 5th International Power Electronics and Motion Control Conference, DOI: 10.1109/IPEMC.2006.4778120
25. Wang, M.-S., Hsieh, M.-F., & Lin, H.-Y. (2018). "Operational Improvement of Interior Permanent Magnet Synchronous Motor Using Fuzzy Field-Weakening Control", *Electronics*, 7(12), 452; https://doi.org/10.3390/electronics7120452
26. Li, M. (2009). "Flux-Weakening Control for Permanent-Magnet Synchronous Motors Based on Z-Source Inverters". Milwaukee: Marquette University thesis, https://epublications.marquette.edu/cgi/viewcontent.cgi?article=1285&context=theses_open
27. Morimoto, S., Takeda, Y., Hirasa, T., & Taniguchi, K. (1990). "Expansion of Operating Limits for Permanent Magnet Motor by Current Vector Control", *IEEE Transactions on Ind. Appl.*, Vol. 26 Issue 5, DOI: 10.1109/28.60058
28. Huang, S., Chen, Z., Huang, K., & Gao, J. (2010). "Maximum Torque Per Ampere and Flux-weakening Control for PMSM Based on Curve Fitting", 2010 IEEE Vehicle Power and Propulsion Conference, DOI: 10.1109/VPPC.2010.5729024
29. Lu, D., & Kar, N. C. (2010). "A Review of Flux-weakening Control in Permanent Magnet Synchronous Machines", 2010 IEEE Vehicle Power and Propulsion Conference, DOI: 10.1109/VPPC.2010.5728986
30. Bolognani, S., Petrella, R., Calligaro, S., & Pogni, F. (2011). "Flux-Weakening in IPM Motor Drives: Comparison of State-of-Art Algorithms and a Novel Proposal for Controller Design", http://www.diegm.uniud.it/petrella/Azionamenti%20Elettrici%20I/Full%20paper%20-%20EPE%202011.pdf
31. Eldeeb, H., Hackl, C. M., Horlbeck, L., & Kullick, J. (2018). "A unified theory for optimal feedforward torque control of anisotropic synchronous machines", *International Journal of Control*, Vol. 91 - Issue 10, https://doi.org/10.1080/00207179.2017.1338359

32. C.M. Hackl, J. Kullick and N. Monzen, "Generic loss minimization for nonlinear synchronous machines by analytical computation of optimal reference currents considering copper and iron losses", accepted for publication in *Proceedings of the 22nd IEEE International Conference on Industrial Technology (ICIT), Special Session: Electrical Drives for Electric Mobility and Green Energy*, 2021.

2

Comparative Analyses of the Response of Core Temperature of a Lithium Ion Battery under Various Drive Cycles

Sumukh Surya[1]* and Vineeth Patil[2]

[1]*Bosch Global Software Technologies, Bangalore, India*
[2]*Dept. of Electrical and Electronics, MIT, Manipal, India*

Abstract

Core temperature (T_c) estimation, State of Charge (SOC) and State of Health (SOH) are the key algorithms in any Battery Management System (BMS). During high-speed operation of the Electric Vehicle (EV), large current is drawn from the battery due to which T_c reaches a higher value than surface temperature (T_s). This leads to thermal runaway causing degradation in SOH, resulting in increased internal resistance and/or reduction in capacity. In the present work, T_c for a Lithium (Li) ion cell under various drive cycles is estimated using Kalman Filter. Based on governing coupled heat transfer equations, T_c is modeled using measured T_s and ambient temperature (T_{amb}). Simulation was carried out using MATLAB/Simulink software using "Commonly Used Blocks". The standard drive cycle pattern was chosen from the "Vehicle Dynamics Blockset" available in the software. Using the dynamic equation of the EV, the velocity profile was converted to current and was fed as one of the inputs to the thermal model. A simple cell model consisting of 1 RC pair was used for obtaining the Open Circuit Voltage (OCV) and terminal voltage used for T_c estimation. The OCV – SOC values were modeled as a 1-D lookup table in Simulink. It was observed that T_c and T_s closely followed the current pattern and $T_c > T_s$ during current discharge. Among all the drive cycles considered, Highway Fuel Economy Test (HWFET) showed maximum T_c of 300.9K, thereby highlighting the importance of simulation studies before hardware development.

**Corresponding author*: sumukhsurya@gmail.com

Pedram Asef, Sanjeevikumar Padmanaban, and Andrew Lapthorn (eds.) Modern Automotive Electrical Systems, (55–74)

Keywords: Battery management system (BMS), drive cycle, Li ion cell, MATLAB/Simulink, thermal model, thermal management

2.1 Introduction

EVs are considered superior to conventional vehicles as they cause less air pollution. Electric Vehicle/Hybrid Electric Vehicles (HEV) generally use energy storage devices like Batteries, Fuel Cells and Ultra Capacitors. Amongst various energy sources, Lithium (Li) ion battery packs are preferred as they possess high energy and specific density. However, they are highly sensitive to temperature (generally high temperature reduces battery life). Therefore, they must be operated under certain range of temperature for better performance and life [1, 2]. Figure 2.1 depicts the safe operating region for a Li ion cell. With the increase in temperature of the battery, State of Health (SOH) of the battery decreases, which results in power and capacity fade.

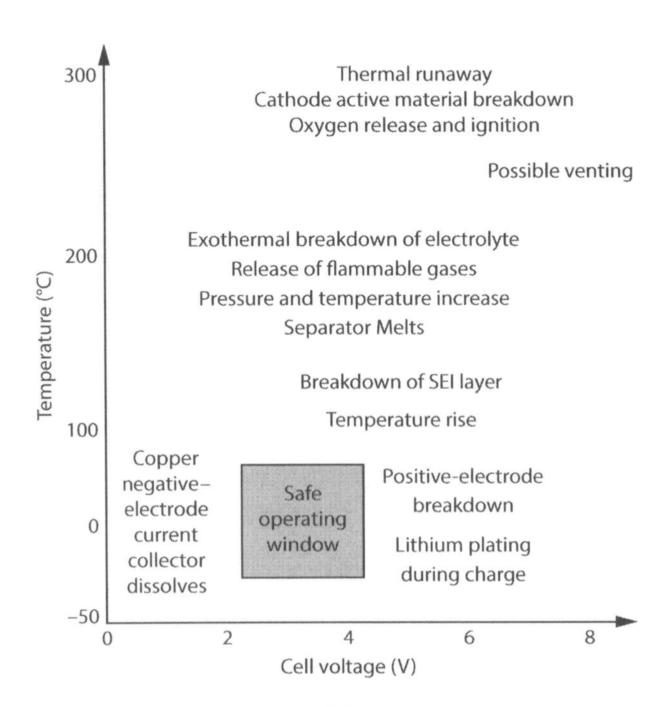

Figure 2.1 Temperature vs. cell voltage (V) [1].

T_c of the cell increases during discharge and can exceed the threshold if not kept under control. In [3], the effect of fast discharge of a Li ion battery on its T_c was estimated for different C rates. Amongst the various C rates considered, maximum T_c was for constant 2C discharge (160°C) for a run time of 50s. Hence, to design an effective thermal management emphasis on T_c is mandatory. In [4], a second order thermal model was considered for the estimation of T_c using heat transfer equations. The equations were mathematically modeled using MATLAB/Simulink by selecting a proper step time shown in [5]. A Kalman filter was implemented using "Commonly Used Blocks" to estimate T_c from known T_s and T_{amb}. Estimated T_c and the actual T_c matched with a maximum error of 5°C. The batteries considered were Lead acid batteries of 60Ah and 68Ah.

In [6], T_c for a Li ion 18650 cell was estimated using Kalman Filter (KF). It is shown that the thermal parameter C_s had no role during T_c estimation. However, this parameter played a crucial role in estimating T_s from known T_c. The value of C_s (Heat Capacity at the core) has to be smaller than C_c (Heat Capacity at the surface) for minimal error in estimation.

The output voltage of battery is generally fed to a Switch Mode Power Converter (SMPS) for obtaining stabilized voltage. Some of the typical converters used in charging systems are Cuk and SEPIC. In [7, 8], Cuk and SEPIC operating in Continuous Conduction Mode (CCM) and Discontinuous Conduction Mode (DCM) are modeled and their dynamics are captured. These converters were unstable in the open loop configuration for constant voltage (G_{vd}) CCM operation [6]. However, converters showed high stability when operated in DCM operation. Generalized circuit averaging technique was used to model the converters for G_{vd} operation.

Modeling of DC – DC converters is extremely important in charging systems for designing passive elements. The mathematical model developed must provide information on the transient and steady state behavior. The equations were modeled using volt-sec and amp-sec balance for ideal DC – DC converters. Such models were developed and simulated using MATLAB/Simulink [8, 9]. The topology involves Buck, Boost, Cuk, SEPIC and Isolated converters like Flyback and Forward. Since the developed models used "Commonly Used Blocks", the C code can be easily generated. In [10], different modeling techniques of DC – DC converters and the extraction of DC transfer function were shown.

One of the common problems associated with AC charging systems is achieving Unity Power Factor (UPF). Popular current control techniques like Peak Current Mode (PCM) and Average Current Control (ACC) are employed to achieve UPF. In ACC, ratio of the inductor current to the duty ratio (G_{id}) is modeled and a control algorithm is developed. The transfer function of G_{id} is mathematically modeled using various control techniques like Small Signal Analysis and State Space Averaging (SSA) etc. Although ACC cannot provide quick control, it offers high noise immunity [11]. ACC for Wide Frequency (WF) and High Frequency (HF) non-ideal DC-DC buck converter operating in CCM was presented [12]. The transfer functions for WF operation showed a zero whose frequency was lower than that of corner frequency of the complex poles. In a nutshell, BMS and Power Electronics play a critical model in the design of EV.

For estimating T_c, a second order thermal model was developed. The inputs to the thermal model are (i) Current, (ii) T_{amb}, (iii) T_s, and (iv) OCV, and (v) Terminal Voltage (V_T) which are supplied from a battery. Hence, an efficient battery model has to be modeled for estimating OCV.

In the past, several attempts have been made on development of efficient battery models. In [13], a five RC pair battery model called Enhanced Self Correcting (ESC) model for a Li polymer battery is proposed. It was shown that the battery parameters are independent on the magnitude of discharge current. This type of RC model for a battery was used for pulse charging application. In [14, 15], a two RC pair battery model was proposed for EV application for Li ion polymer batteries. The one RC model consists of R_0 (internal resistance, mΩ), R_p (polarizing resistance, mΩ) and C_p (polarizing capacitance, kF) [16–18]. The standard procedure for obtaining the battery parameters is by performing Pulse Characterization test.

Fast charging is a popular technique used to charge the battery pack quickly without deteriorating the SOH of the battery pack [19]. In [20], an extensive review on extreme fast charging technology is presented. The traditional chargers used were slow chargers due to the service transformer being expensive and causing problems during installation. However, due to the recent developments in the magnetics and Solid-State Transformers (SST), fast charging can be employed. Figures 2.2 and 2.3 show a block diagram representation of slow and fast charging stations.

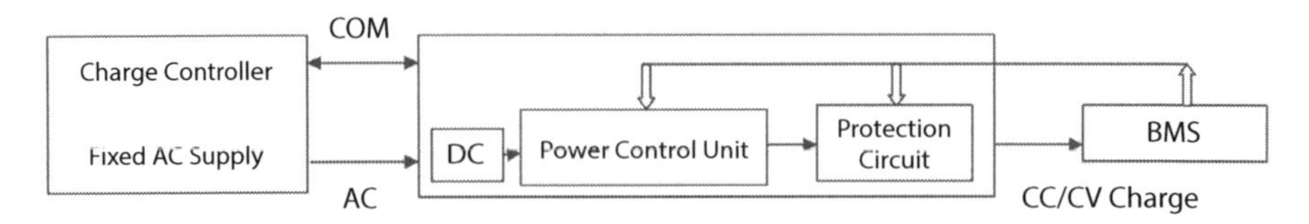

Figure 2.2 Schematic of a slow charger.

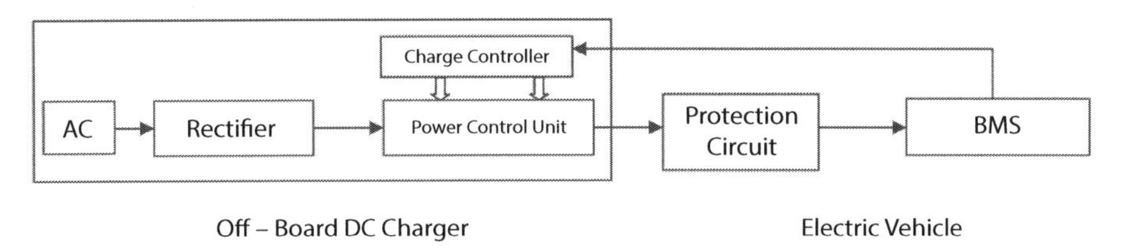

Figure 2.3 Block diagram of a fast charger [20].

Table 2.1 shows commercially available chargers for EV considering the ratings, type and capacity of the battery and the range of EV.

The charging levels, configuration and standards are shown in Table 2.2. Figure 2.4 shows the charger connections.

The charger connections are diversified as Type 1, Type 2, etc., which are explained in detail in [20]. A few examples of DC connectors are CCS Combo 1, 2 and CHAdeMo. Different types of fast charging are (1) AC charging (2) DC Charging. DC charging can be further classified as Uni-polar and Bi-polar DC-bus. Table 2.3 shows a comparison between AC and DC charging.

In [21], a closed loop control on Constant Temperature – Constant Voltage (CT-CV) is performed on a Li ion 18650 cell. CT - CV charging

Table 2.1 Specifications of EV.

EV model	Motor rating (kW)	Motor	Capacity (kWh)	Range (mil)
Smart FortoWo ED	55	PMSM	17.6	58
Hyundai Ioniq Elec.	88	PMSM	28	124
Mahindra Reva	35	IM	16	75
Kai Soul EV	81	PMSM	30	110
Renault Zoe	80	PMSM	41	250
Tesla Model 3	192	PMSM	75	220
Tesla Model S 70D BEV	100	IM	100	240

Table 2.2 Levels of charging.

Level	Range of voltage	P_{max} (kW)	Time for charging
1 (Slow)	120V AC	3.7	10-15 hrs
2 (Slow)	220V AC	37-22	3.5-7 hrs
3 (Fast)	3Phase 480V AC, 200-600V DC	22-43.5, < 200	10-30 min

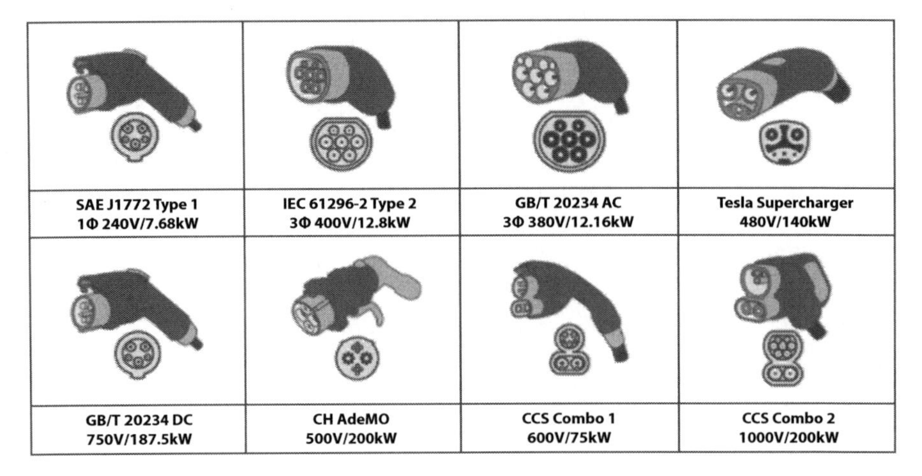

Figure 2.4 Charger connection [20].

Table 2.3 Comparison between AC and DC coupled system.

Parameter	Coupling to AC	Coupling to DC
Technical Maturity	High	Low
Availability	High	Low
Complexity of Protective Devices	Low	High
Conversion Stages	High	Low
Efficiency	Low	High
Control Complexity	High	Low
Cost	High	Low

provides faster charging than the traditional Constant Current – Constant Voltage (CC - CV) charging [21, 22]. In this method, a certain amount of current is pushed into the cell. T_s corresponding to the current is measured using thermo-couples placed at appropriate locations. If T_s crosses the threshold limit, current is reduced to a minimum value to maintain constant voltage.

The SOH can get degraded due to which the capacity gets reduced over long usage and/or the internal resistance of the battery (R_0) increases. The former is referred to as Capacity Fade and the latter as Power Fade. This is fundamentally due to the inherent chemical reactions and the structural deterioration. In practical scenario, Li ion is lost due to the side reaction. This phenomenon is called Loss of Li Inventory (LLI) [18]. Table 2.4 shows the aging of the cell which occurs due to various reasons.

In the present work, T_c is estimated based on measured T_s and T_{amb} using governing heat-transfer equations. The discharge current is applied to the thermal model for various drive cycle data viz., US06 (Supplemental Federal Test Procedure), HWFET (Highway Fuel Economy Test) and UDDS (Urban Dynamometer Driving Schedule). The drive cycle profile (velocity) was available on MATLAB/Simulink under Vehicle Dynamics Blockset. Comparative analysis on T_c for different drives is carried out.

Table 2.4 Aging of the cell due to composite electrode materials.

Cause	Effect	Effect	Enhanced by
Gas evolution, Graphite Exfoliation	Loss of Li	Capacity Fade	Overcharge
Contact loss between particles due to volume changes	Loss of Li	Capacity Fade	High rate, Low cell SOC
Decomposition of binder	Li loss, loss of mechanical stability	Capacity Fade	High SOC, high temperatures
Current collector corrosion	Rise in impedance	Power Fade	Over discharge, low cell SOC

2.2 Thermal Modeling

Figure 2.5 shows a second order thermal model having two capacitors C_c and C_s. Since the circuit contains two energy storage elements it is considered as a Second Order Thermal Model. Comparison between the first order and second order thermal models are shown in [23]. It was shown that in order to have precise and accurate T_c value, second order thermal model is preferred.

Equations (2.1) and (2.2) provide T_c and T_s for known values of I (A), terminal voltage V_T (V), T_{amb} (K) and OCV (V) by applying Kirchhoff's Current and Voltage law.

$$C_c \frac{dT_c}{dt} = Q + (T_s - T_c)/R_c \tag{2.1}$$

$$C_s \frac{dT_s}{dt} = (T_{amb} - T_s)/R_u - (T_s - T_c)/R_c \tag{2.2}$$

$$Q = I(V_T - OCV) \tag{2.3}$$

where, C_c and C_s are the heat capacities of the core and surface (J/K), respectively, and R_c and R_u are the convective resistances between the core and surface and surface and ambient (K/W) respectively.

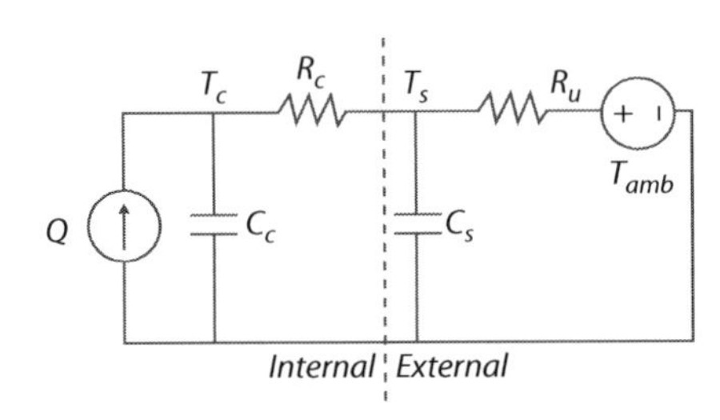

Figure 2.5 Second order thermal model.

2.3 Methodology

The plant model of Li ion cell containing two resistor – capacitor (RC) pairs for a Li ion polymer chemistry was modeled using MATLAB/Simulink. The code used optimization and parallel computing tool boxes for computing RC values [14]. T_{amb} and T_s were fixed at 298K and 299K, respectively. Based on equations (2.1), (2.2) and (2.3), KF was developed based on [4, 6]. The process noise P and the measurement noise levels were appropriately chosen. For computing the SOC of a Li ion cell, Coulomb Counting (CC) method was employed.

The current required to discharge the battery was supplied from the drive cycle block. Multiple drive cycles available under the Vehicle Dynamics Blockset are FTP75, US06, HWFET, UDDS, etc. The velocity in m/s was converted to current (A) using the equations shown below

$$F = F_D + F_R + F_c \tag{2.4}$$

$$F_D = 0.5\rho C_d A V^2 \tag{2.5}$$

$$F_R = C_R mg \tag{2.6}$$

$$F_C = mgSin\theta \tag{2.7}$$

F_d is aerodynamic drag, F_R is rolling friction, F_c is climbing force, ρ is density (kg/m^3), C_d is drag co-efficient, A is frontal area (m^2), V is velocity (m/s), C_R is rolling friction co-efficient, m is mass of the EV, g is acceleration due to gravity (m/s^2), θ is the elevation angle (Degree).

If the velocity of the air is in the direction of the EV

$$F_D = 0.5\rho C_d A(V + V_{air})^2 \tag{2.8}$$

Similarly, if the velocity of the air is against the direction of EV,

$$F_D = 0.5\rho C_d A(V - V_{air})^2 \quad (2.9)$$

$$P = FV \quad (2.10)$$

P is power (kW), V is voltage (V) supplied to the motor

Hence, $$P = 0.5\rho C_d A V^3 + C_R mgV + mgVSin\theta \quad (2.11)$$

Equation (2.11) can be reduced in the form shown in (2.12)

$$P = AV + BV^2 + CV^3 \quad (2.12)$$

C = C_d, A is the rolling resistance and B is climbing resistance and are specific to the EV.

Table 2.5 shows the specifications of the thermal model.

Table 2.5 Thermal model specifications.

SL. no.	Parameter	Value
1	R_u	0.26 K/W
2	R_c	0.864 K/W
3	C_c	1067 J/K
4	C_s	545.3 J/K
5	A	177.2
6	B	1.445
7	C	0.354

2.4 Simulation Results

The variation of R_0 is shown in Figure 2.6. It can be seen that as the time decreases, SOC decreases and the internal resistance also decreases. However, at around 900s, R_0 increases as the SOC further decreases. This parameter estimation plays a very important role in estimation of power fade.

The variation of C_2 is shown in Figure 2.7. It can be observed that the C_2 increases with the drop in SOC. For a short period of time (low SOC), the C_2 is constant. Later, the value monotonically increases.

From Figures 2.5 and 2.6 it can be concluded that the battery parameters have strong dependence on T_{amb} and SOC.

Figures 2.8 and 2.9 show the velocity (m/s) and the calculated current, respectively, for US06 drive cycle. This drive cycle was developed to overcome the limitations in FTP75 drive cycle. Figure 2.10 shows a plot of measured T_s and estimated T_c. It is observed that as the velocity of the EV increases, the current drawn from battery and T_c increases gradually. The current reached a maximum value of about 102A whereas rise in T_c and T_s was slow due to thermal inertia. This inertia is strongly governed by the thermal parameters.

Figures 2.11 and 2.12 show the velocity and current, respectively, for HWFET drive cycle. HWFET was developed by the US Environmental Protection Agency for light-duty vehicles for estimating the fuel economy.

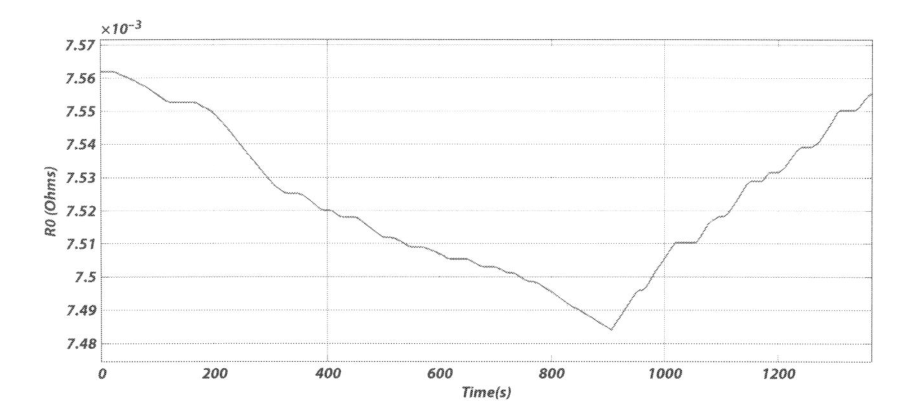

Figure 2.6 R_0 vs time for discharge current.

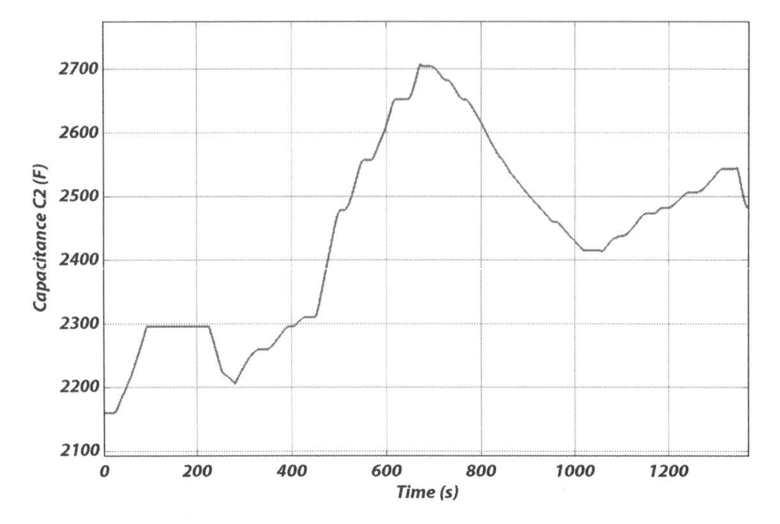

Figure 2.7 C_2 vs time for discharge current.

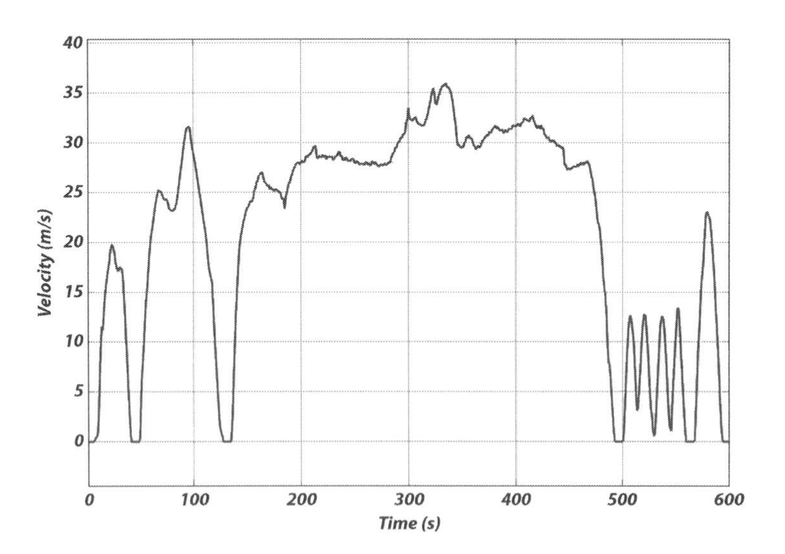

Figure 2.8 US06 drive cycle.

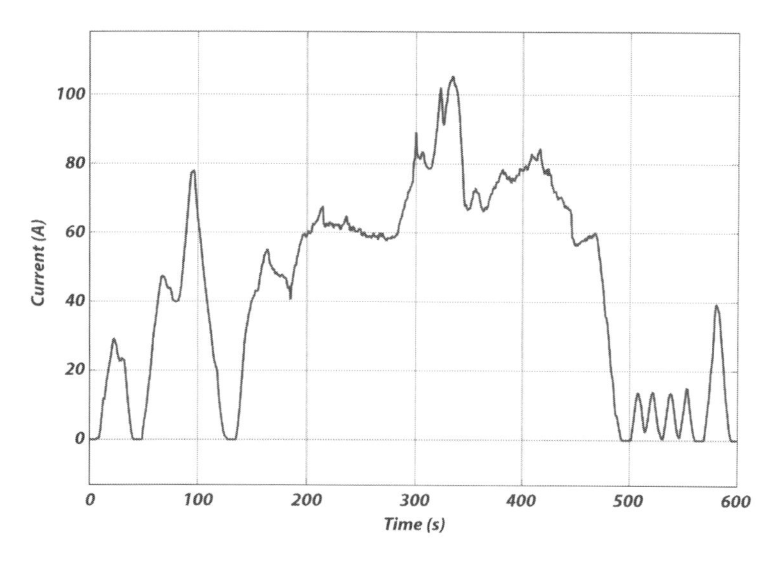

Figure 2.9 Current from drive cycle.

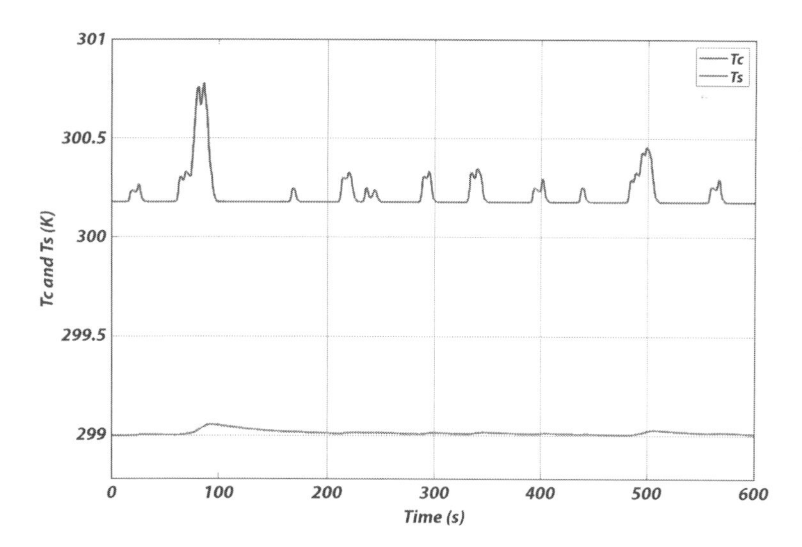

Figure 2.10 T_c and T_s vs. time.

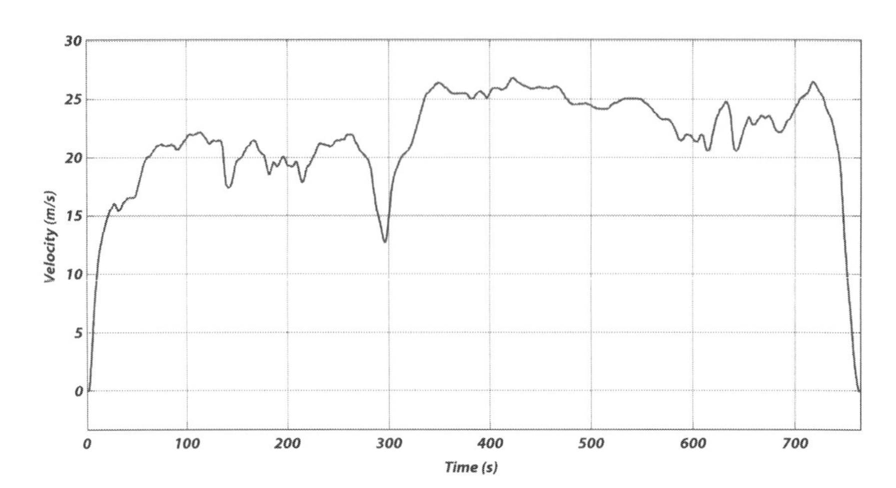

Figure 2.11 HWFET drive cycle.

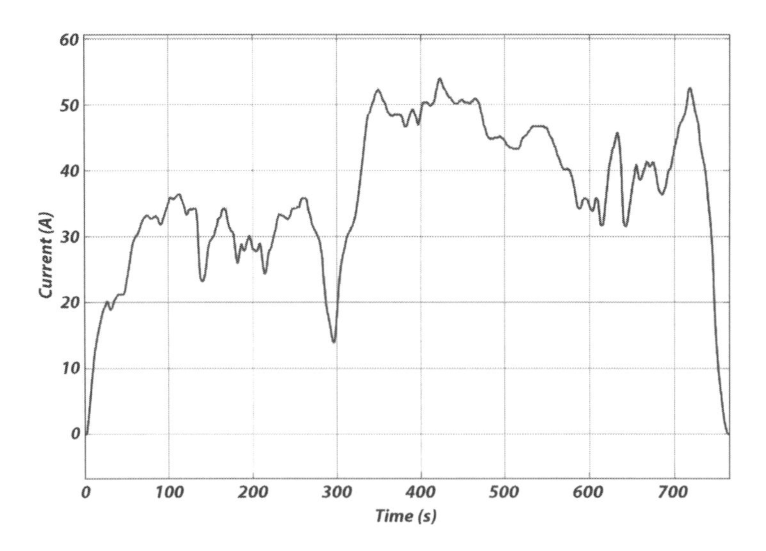

Figure 2.12 Current from drive cycle.

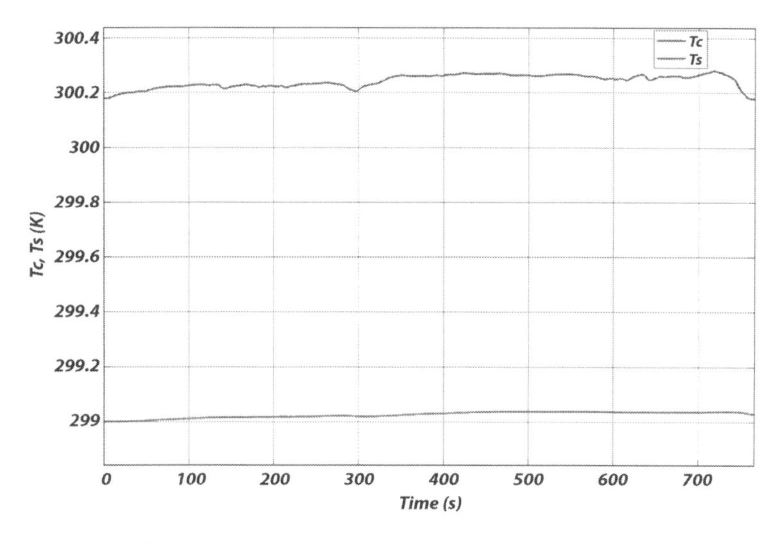

Figure 2.13 T_c and T_s vs. time.

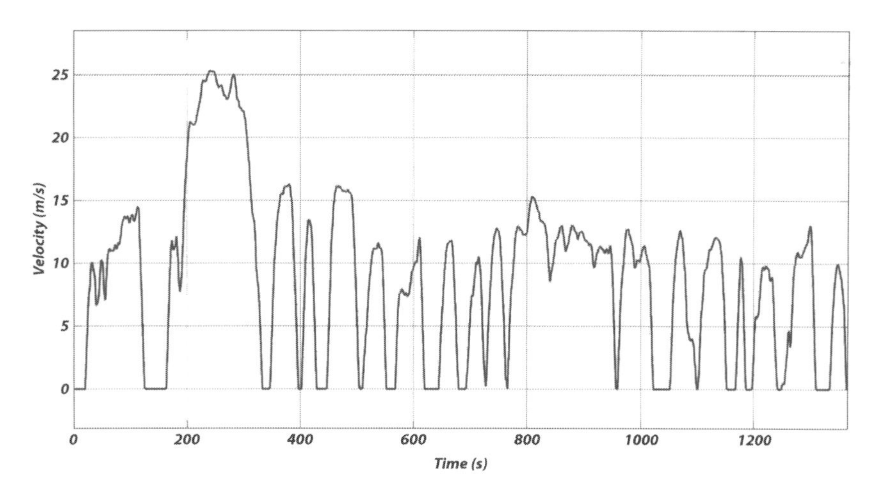

Figure 2.14 UDDS drive cycle.

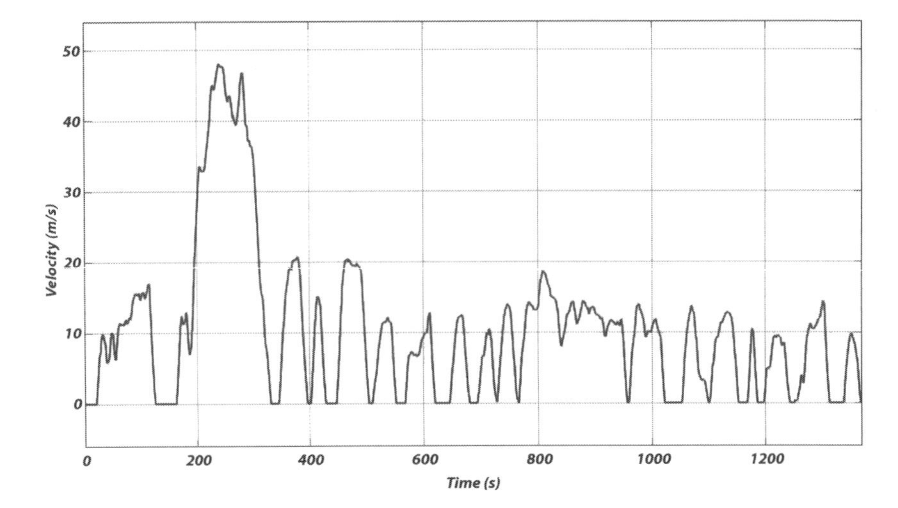

Figure 2.15 Current from drive cycle.

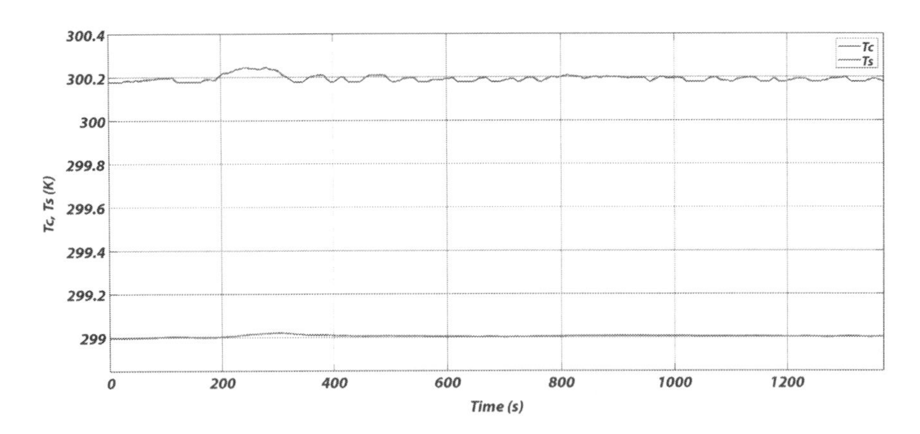

Figure 2.16 T_c and T_s vs. time.

Table 2.6 Comparison of T_c and T_s under different drive cycles.

Drive cycle	T_{cmax} (K)	T_{smax} (K)
US06	300.8	299.25
HWFET	300.3	299.25
UDDS	300.22	299

Figure 2.13 shows a plot of measured T_s and estimated T_c. The current increased from 0 to 120s due to which T_c also increased. Around 130s, the current waveform showed a dip due to which T_c also lowered. It can be observed that the temperature rise is higher in US06 than HWFET as the magnitude of velocity is larger.

Figures 2.14 and 2.15 show the velocity and current profiles respectively for UDDS drive cycle. UDDS (Urban Dynamometer Driving Scheme) was developed by the US Environmental Protection Agency for light-duty vehicles for fuel economy estimation. It also referred to as FTP75. Figure 2.16 shows a plot of measured T_s and estimated T_c. Observations similar to those in Figures 2.8–2.10 were made.

Table 2.6 shows T_c estimate for different drive cycles until the SOC was 5%. It is noted that the US06 showed highest T_c compared to other drive cycles. This is due to the larger magnitude in the velocity profile.

2.5 Conclusions

T_c for a 94Ah Li ion battery was estimated using heat transfer equations. The equations were modeled using appropriate step time using MATLAB/ Simulink. Since the KF based on the second order thermal model was developed using "Commonly Used Blocks", the C code can be easily generated with no dependence on additional toolboxes. T_c estimation for US06, HWFET and UDDS drive cycles was estimated. It is observed that $T_c > T_s$ and the magnitude can be larger if the C rate increases. The temperatures closely follow the current pattern. However, the sudden increase in current does not cause rapid change in the temperatures due to thermal inertia. The thermal inertia is caused due to the large value of the thermal constants between the surface and core namely R_c, R_u, C_c and C_s. Among the various drive cycles considered, US06 drive cycle showed T_{cmax} and T_{smax} were about 300.8K and 299.25K.

References

1. Plett, Gregory L. *Battery management systems*, Volume I: *Battery modeling*. Artech House, 2015.
2. Surya, S.; Williamson, S. "Energy Storage Devices and Front-End Converter Topologies for Electric Vehicle Application." Accepted for Publication in *E-Mobility—A New Era in Automotive Technology*; Springer: Berlin/ Heidelberg, Germany, 2021.

3. Surya, Sumukh, and M. N. Arjun. "Effect of fast discharge of a battery on its core temperature." 2020 International Conference on Futuristic Technologies in Control Systems & Renewable Energy (ICFCR). IEEE, 2020.
4. Surya, Sumukh, *et al.* "Development of thermal model for estimation of core temperature of batteries." *International Journal of Emerging Electric Power Systems* 21.4 (2020).
5. Surya, Sumukh, and Vineeth Patil. "Cuk Converter as an Efficient Driver for LED." 2019 4th International Conference on Electrical, Electronics, Communication, Computer Technologies and Optimization Techniques (ICEECCOT). IEEE, 2019.
6. Surya, Sumukh, Vinicius Marcis, and Sheldon Williamson. "Core Temperature Estimation for a Lithium ion 18650 Cell." *Energies* 14.1 (2021): 87.
7. Surya, Sumukh, and Sheldon Williamson. "Generalized Circuit Averaging Technique for Two-Switch PWM DC-DC Converters in CCM." *Electronics* 10.4 (2021): 392.
8. Surya, Sumukh, JanamejayaChannegowda, and Kali Naraharisetti. "Generalized Circuit Averaging Technique for Two Switch DC-DC Converters." *arXiv preprint arXiv:2012.12724* (2020).
9. Surya, Sumukh, and M. N. Arjun. "Mathematical Modeling of Power Electronic Converters." *SN Computer Science* 2.4 (2021): 1-9.
10. Surya, Sumukh. "Mathematical Modeling of DC-DC Converters and Li Ion Battery Using MATLAB/Simulink." *Electric Vehicles and the Future of Energy Efficient Transportation*. IGI Global, 2021. 104-143.
11. Surya, Sumukh, and Sheldon Williamson. "Modeling of Average Current in Ideal and Non-Ideal Boost and Synchronous Boost Converters." *Energies* 14.16 (2021): 5158.
12. Kondrath, Nisha, and Marian K. Kazimierczuk. "Comparison of wide- and high-frequency duty-ratio-to-inductor-current transfer functions of DC–DC PWM buck converter in CCM." *IEEE Transactions on Industrial Electronics* 59.1 (2011): 641-643.
13. Surya, Sumukh, *et al.* "Accurate Battery Modeling Based on Pulse Charging using MATLAB/Simulink." *2020 IEEE International Conference on Power Electronics, Drives and Energy Systems (PEDES)*. IEEE, 2020.
14. Lam, Long, Pavol Bauer, and Erik Kelder. "A practical circuit-based model for Li-ion battery cells in electric vehicle applications." *2011 IEEE 33rd International Telecommunications Energy Conference (INTELEC)*. IEEE, 2011.
15. Chen, Min, and Gabriel A. Rincon-Mora. "Accurate electrical battery model capable of predicting runtime and IV performance." *IEEE Transactions on Energy Conversion* 21.2 (2006): 504-511.
16. Omariba, Zachary Bosire, *et al.* "Parameter Identification and State Estimation of Lithium-Ion Batteries for Electric Vehicles with Vibration and Temperature Dynamics." *World Electric Vehicle Journal* 11.3 (2020): 50.

17. Plett, Gregory L. *Battery management systems, Volume II: Equivalent-circuit methods*. Artech House, 2015.
18. Zheng, Yongliang, Feng He, and Wenliang Wang. "A method to identify lithium battery parameters and estimate SOC based on different temperatures and driving conditions." *Electronics* 8.12 (2019): 1391.
19. Surya, S.; Rao, V.; Williamson, S.S. "Comprehensive Review on Smart Techniques for Estimation of State of Health for Battery Management System Application." *Energies* 2021, *14*, 4617. https://doi.org/10.3390/en14154617.
20. Ronanki, Deepak, Apoorva Kelkar, and Sheldon S. Williamson. "Extreme fast charging technology—Prospects to enhance sustainable electric transportation." *Energies* 12.19 (2019): 3721.
21. Patnaik, Lalit, A. V. J. S. Praneeth, and Sheldon S. Williamson. "A closed-loop constant-temperature constant-voltage charging technique to reduce charge time of lithium-ion batteries." *IEEE Transactions on Industrial Electronics* 66.2 (2018): 1059-1067.
22. Marcis, Vinicius Albanas, *et al.* "Analysis of CT-CV charging technique for lithium-ion and NCM 18650 cells." *2020 IEEE International Conference on Power Electronics, Smart Grid and Renewable Energy (PESGRE 2020)*. IEEE, 2020.
23. Surya, Sumukh, Akash Samanta, and Sheldon Williamson. "Smart Core and Surface Temperature Estimation Techniques for Health-conscious Lithium-ion Battery Management Systems: A Model-to-Model Comparison." (2021).

3

Classification and Assessment of Energy Storage Systems for Electrified Vehicle Applications: Modelling, Challenges, and Recent Developments

Seyed Ehsan Ahmadi[1]* and Sina Delpasand[2]

[1]Department of Electrical Engineering, University of Kurdistan, Sanandaj, Iran
[2]Faculty of New Sciences and Technologies, University of Tehran, Tehran, Iran

Abstract

The electric vehicle (EV) technology resolves the need to decrease greenhouse gas emissions. The principle of EVs concentrates on the application of alternative energy resources. However, EV systems presently meet several issues in energy storage systems (ESSs) concerning their size, safety, cost, and general management challenges. Furthermore, the hybridization of ESSs with developed power electronic technologies has a major effect on optimal energy usage to manage modern EV technologies. This chapter comprehensively reviews ESS technologies, classifications, characteristics, and evaluation procedures with pros and cons for EV applications. Besides, this chapter addresses diverse classifications of ESS based on their composition materials, energy formations, and approaches on power delivery over its potential and performances indicated within their life expectancies. This chapter also presents the modeling and controls of ESSs in EV applications. Algorithms of the battery system play an essential role in EVs due to the direct effects on the overall fuel economy, drivability, and safety of an EV. However, because of the intricacy of electrochemical reactions, dynamics, and availability of main variable measurements, EV systems are dealing with technical issues in the advancement of the required algorithms for EVs. In this chapter, the state of charge determination algorithms and the corresponding methodical issues are also reviewed. Moreover, the power capability and state of life algorithms are comprehensively addressed. This chapter highlights various factors, issues, and

**Corresponding author*: ehsan.ahmadi@uok.ac.ir

Pedram Asef, Sanjeevikumar Padmanaban, and Andrew Lapthorn (eds.) Modern Automotive Electrical Systems, (75–148)

challenges for sustainable improvement and future extensions of ESS technologies in state-of-the-art EV applications. Therefore, this chapter will expand the effort toward the improvement of efficient and economic ESSs with an extended lifetime for future EV applications.

***Keywords*:** Energy storage system, electrified vehicle, state of charge determination, life prediction

3.1 Introduction

In recent decades, renewable energy sources (RES) have been supplying separated power users and electrical networks with energy storage systems (ESSs) [1, 2]. In this regard, electric vehicles (EVs) are developing technologies with ESS as a replacement for fossil fuel devices. The EVs are employed to prevent the employment of fossil fuel devices and decrease the released CO_2. Therefore, energy-efficient ESSs are essential to driving EVs conveniently [3]. To ensure some necessities of EVs, the ESSs are employed in cooperation to enable greater discharge time with self-reliance. Advanced ESSs are presenting new facilities in EVs. The algorithms associated with the ESSs play a major role in EV implementations. Establishing advanced facilities in the EVs necessitates managing energy resources, selecting proper ESSs, balancing the charge of the storage cell, and impeding abnormality [4]. EVs are highly reliant on existing energy storage technologies, such as battery energy storage, fuel cells, and ultracapacitors for power. Therefore, EVs need to be charged from the utility power grid [5]. Other energy storage technologies including flywheel, superconducting magnetic coils, and hybrid ESSs can be also applied in EV energizing applications. Besides, the saved energy in an EV can be transferred to the grid as a vehicle-to-grid (V2G) facility. The electric energy stored in the ESSs is utilized to drive the electric motor and its accessories, as well as the primary electric systems of the EV to operate [6]. From basic current, voltage, and temperature measurements that determine the values of functional variables, to the state of charge (SOC) estimation, power capacity prediction, and the state of life of battery energy storage, battery algorithms have influenced many factors of an EV system. The SOC of a battery is described as a percentage of the obtainable value of energy over its highest capacity. As the robustness of SOC estimation plays an important role in EV function, various SOC estimation methods have been established. According to the obligations for output voltage and power potential for the EV, a battery pack is set up by several cells connected in series or parallel, or both. The framework of the electrical circuit model is commonly figured out offline according to the test data. Furthermore, effectively controlling the state of health of the ESS greatly enhances the reliability of an EV [7].

From the literature, the majority of the ESS review papers consider the technologies applied for storing secondary energy forms. Guney and Tepe [8] provide a comprehensive description of ESSs with thorough classification, characteristics, advantages, environmental effects, and operation possibilities with application variations. Aneke and Wang [9] discuss the concept of ESS, the diverse technologies for storing energy with a focus on the storage of secondary forms of energy such as electricity and heat, and a thorough analysis of different ESS projects throughout the world. Das *et al.* [10] provide a comprehensive review of the present situation of the EV market, required standards, charging/discharging infrastructures, and the effects of EV charging on the power network. Sharma *et al.* [11] present a comprehensive review of diverse components and ESSs applied in EVs. The main objective of this paper is battery technologies as the key factor in making EVs more eco-friendly, economic, and drives the EVs into utilization in daily life. Kouchachvili *et al.* [12] review the recent studies dedicated to the diverse battery/supercapacitor hybrid systems in EV application. Machlev *et al.* [13] also review recent studies allied with the optimal control of ESSs with classic and advanced methods. Balali and Stegen [3] present a review of ESSs for light-duty vehicles and highlight the major characteristics of EVs according to the environmental perspective, power train structure, and cost. Xiao *et al.* [14] review the current improvements of EVs relating to energy management, powertrain structure, and vehicle performance. Zhang *et al.* [15] present a hybrid ESS overview focusing on battery-supercapacitor hybrids, reviewing diverse aspects in smart grid and EV applications. Koohi-Fayegh and Rosen [6] review various ESS technologies, ESS types, and categorizations. Ibrahim and Jiang [16] provide a comprehensive review containing the energy management of EVs regarding the storage and consumption systems. They also introduce diverse ESSs applied in EVs with more elaborate details on Li-ion batteries. Hannan *et al.* [17] present a comprehensive review of technologies of ESSs, their classifications, and evaluation processes with advantages and disadvantages for EV utilization.

The previous works are often restricted concerning the covered ESS types. For instance, some reviews only focus on ESS types for a specific application. Other reviews only focus on electrical ESSs without investigating other ESS types such as thermal or hydrogen-based ESSs. It is vital that more general reviews covering all ESS types are conducted to present higher insights on their variations, potential integration opportunities, and required policy evolution. On the other hand, the methodologies of calculating the SoC and life prediction of batteries are not comprehensively reviewed in previous works. The main contribution of this chapter can be described as the thorough investigation of various ESS technologies, SOC determination, and power capability, and state of life algorithms for EV implementations. The outcomes provide a practical idea for scientific researchers and EV manufacturers on the current ESS technologies and their improvements for future applications. A detailed combination of current ESSs could impact future research and improvement of the ESS hybridizations with corresponding aspects for effective energy usage in EV applications. The review on ESS technologies and the corresponding mathematical algorithms for EV application might introduce a concrete concept on the drive train architectures of the EV in such a way that the research might concentrate on the proper utilization of the ESSs and their control strategies in EV. This chapter also focuses on issues and challenges of the current ESS technologies. Such issues could lead academics and manufacturers to evaluate capabilities of the adjustment, development, and alternation in ESS improvements, and to enable technologies for overcoming issues toward sustainable ESS for EV application.

This chapter aims to model ESSs for EVs and explain the corresponding technologies. Initially, the background of EVs is introduced in Section 3.2. In this section, EV classifications, charging/discharging strategies, and classification of ESSs in EVs mode are addressed. Then, the modeling of ESSs applied in EVs is comprehensively reviewed in Section 3.3. In this section, the mechanical, electrochemical, chemical, electrical, thermal, hybrid ESSs are properly investigated, as well as the electrical and thermal behavior modeling and SOC calculation of the battery storage system as the most common ESS in EVs to cover various aspects of this kind of ESS. In Sections 3.4 and 3.5, the characteristics and application of ESSs in EVs are presented, respectively. Later, in Section 3.6, the approaches of calculating the SOC of the battery are reviewed in detail. Then in Section 3.7, the estimation of battery power availability is extensively assessed. Also, in Section 3.8, the life prediction of the battery is studied. For the sake of further analysis, the recent trends, future extensions, and challenges of ESSs in EV implementations are discussed in Section 3.9. The government policy challenges for EVs are also presented in Section 3.10. Finally, the chapter is concluded with several points in Section 3.11.

3.2 Backgrounds

3.2.1 EV Classifications

In general, EVs can be categorized into two large groups, as illustrated in Figure 3.1, including plug-in EVs (PEVs) and non-plug-in EVs, which are mainly hybrid EVs (HEVs). Besides, the PEVs can be further classified into battery EVs (BEVs) and fuel cell EVs (FCEVs). Furthermore, plug-in hybrid EVs (PHEVs) can be also introduced as a subscription to the PEVs and the HEVs groups. The BEVs apply the electricity stored in their battery to drive the electric motor. The amount of energy stored onboard is defined by the battery size. The FCEVs also drive an electric motor. However, an FCEV stores hydrogen gas in a tank instead of recharging a battery. A fuel cell combines hydrogen with the oxygen existing in the air to provide the required electricity. As a result of the chemical reactions, the FCEVs release water from the exhaust pipe. The power of an FCEV is identified by the size of the installed fuel cell. Also, the amount of power that can be stored is identified by the size of the installed hydrogen fuel tank. On the other hand, the PHEVs are generally provided with smaller batteries since they can devise an internal combustion engine, impel the vehicle directly, and recharge the onboard battery performing an increased mobility range compared to the BEVs and FCEVs [18].

Currently, the most prevalent class of EVs on the market are HEVs. Moreover, although the BEVs provide several benefits compared to the HEVs, such as convenient fuel economy, reduced maintenance costs, and lower emissions, they are still deficiently spread, due to their quite

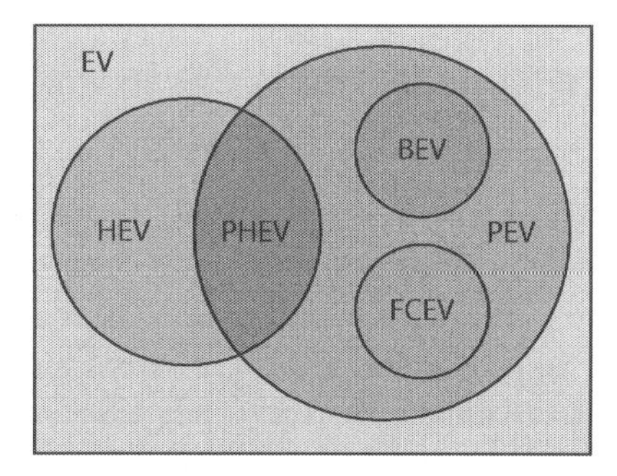

Figure 3.1 EV classifications based on powertrain.

expensive onboard batteries which provide a lower mobility range. This issue can be dealt with by applying proper charging strategies of PEVs according to their huge energy potential, which strives to minimize the PEV operating costs, thus, to relatively refund the higher investment costs related to the HEVs [19].

3.2.2 EV Charging/Discharging Strategies

Although the charging of a single EV cannot considerably influence the operation of the power network, a significant number of the EVs will demand suitable charging/discharging strategies as a result of prominent energy flows. The charging/discharging strategy relies on adjusting set most of the located EVs and the corresponding driving requirements [20].

3.2.2.1 Uncontrolled Charge and Discharge Strategies

In uncontrolled charging and discharging strategies, the scheduling of the requested EVs is not planned in such a way that EVs begin to receive the required charge instantly when coupled to the power network during peak and off-peak hours. It can be mentioned that the uncontrolled charging/discharging strategy is extremely complicated. Besides, the network operator does not acquire any consumer data about the operating system in this strategy, which may lead to issues with network stability, network efficiency, power quality, and SOC of the onboard battery. A huge EV charging and discharging have a crucial effect on the network. Accordingly, the storage devices of EVs provide novel potentials for reliable scheduling of RES-based networks. In this sense, the demand modeling of EV charging is needed to investigate the corresponding aspects of power grids and to indicate the charge and discharge controls of EVs [10].

3.2.2.2 Controlled Charge and Discharge Strategies

Recently, the controlled strategy of EV charging and discharging has attracted increased attraction. This coordinated strategy can be promptly adopted and controlled by the operator, who schedules the charging and discharging periods to prevent problems related to the power quality while ensuring the driving requirements of EV owners and assuring financial or technical objectives. Along with the types of controlling parameters, the developed charging/discharging structure can be categorized into indirect control, smart control, and bi-directional control strategies [10].

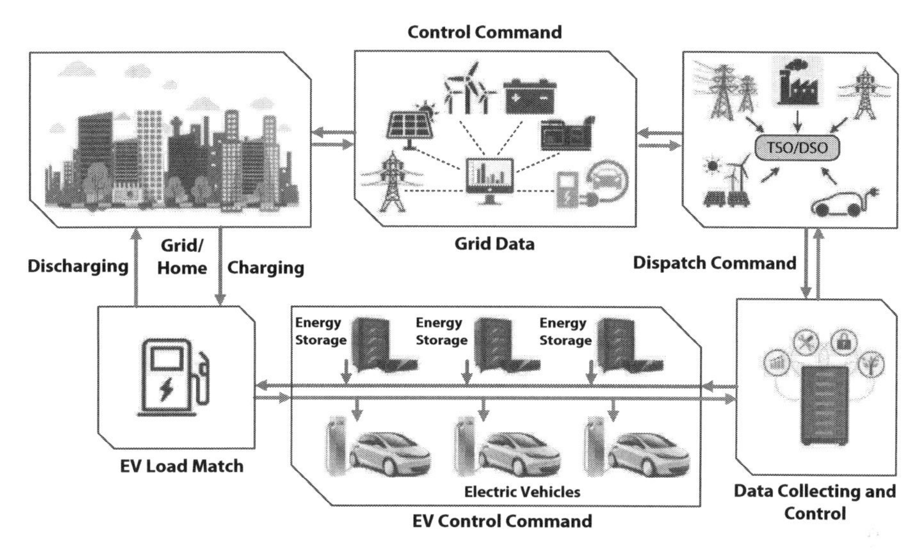

Figure 3.2 Regulated EV charging and discharging strategies applying a control center [20].

Figure 3.2 demonstrates a controlled charge/discharge strategy applying a coordinated charge/discharge control center.

3.2.2.3 Wireless Charging of EV

The growth of wireless charging systems for EVs has gained significant momentum over the past decade due to the desire of smart cities to push away from petrol- and diesel-powered vehicles to make cleaner cities, given the drastic urbanization which is occurring in the world, and partly due to EVs are becoming more efficient and cost-competitive. Via wireless charging systems suitably integrated into vehicles, and situated strategically around a city as well as at owners' homes, there should be no requirement to ever plug in their vehicles. These days, plug-in charging is common but, in the future, will change to the domain of high-power fast-charging where necessary. Ideally, EVs can be charged whenever and wherever they are parked and EV owners should not have to worry about the grid connection, which will happen automatically. Instead of physical cable coupling, the wireless (inductive) connection effectively avoids sparking over plugging/unplugging. In addition, wireless charging initiates new possibilities for dynamic charging/charging while driving. Once realized, EVs will no longer be limited by their electric drive range and the demand for battery capacity will be greatly diminished. Today's wireless charging technologies have 90% efficiencies, which is just 1 or 2% less than plug-in systems. The magnetics of the wireless transformer are necessarily split (the primary on the ground and the secondary on the EV), and power is connected using fields that are formed to exist in the gap. The power transfer performance is further enhanced by turning it on only when a vehicle is ready and demanding power [21, 22].

Figure 3.3 illustrates the power flow and data flow of a typical wireless electric vehicle charging system. On the primary side, there are significantly three parts to the power cupboard: AC-DC rectifier, DC-AC inverter, and the compensation unit. The controllable rectifier can correct the power factor, as well as provide auxiliary control by modulating its DC output voltage. The inverter makes a high-frequency alternating current to energize the primary coils, and its switching state should be regulated online according to the load status. The compensation unit is necessarily an impedance matching network, which is used to neutralize the leakage inductance/capacitance for magnetic/electric coupled power transfer, to attain higher efficiency. Moreover, it lowers the voltage-ampere ratings of power electronics significantly with apparent power decreasing. On the secondary side, to lengthen the battery life and increase the optimal output characteristics, multiple measures could be entertained: once secondary coils are excited, secondary compensation unit is engaged to feed constant current, constant voltage, or step power to the batteries. The data flow is used to certify the charging process reliably and safely. In addition, a series

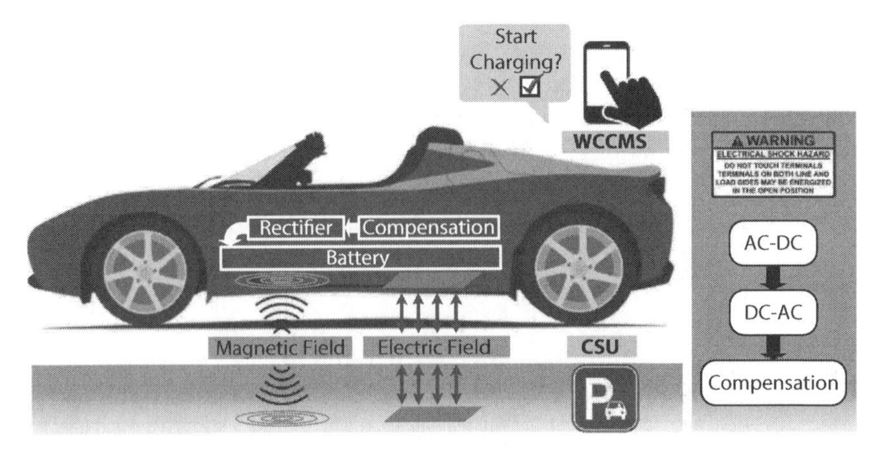

Figure 3.3 The typical wireless EV charging system.

of functions like pairing, information inquiry, and payment all rely on data exchange. Corresponding connection terminals involve a communication service unit (CSU) and wireless charging control and management (WCCMS), which exchange information together from time to time, to ensure the charging process goes forward correctly. If any error happens, the charging process will be stopped immediately. WCCMS is applied for status monitoring of the charging procedure and business operation, a typical example of which is a mobile application [23].

3.2.3 Classification of ESSs in EVs

The classification of ESSs is defined as the employment of energy in a particular form. Generally, the ESS is categorized into mechanical, electrical, chemical, electrochemical, hybrid, and thermal. Figure 3.4 illustrates the detailed classification of common ESS for EV applications. It can be mentioned that FCs, superconducting magnetic coils, UCs, secondary electrochemical batteries, flywheel, and hybrid ESSs are commonly employed in EVs [6, 24].

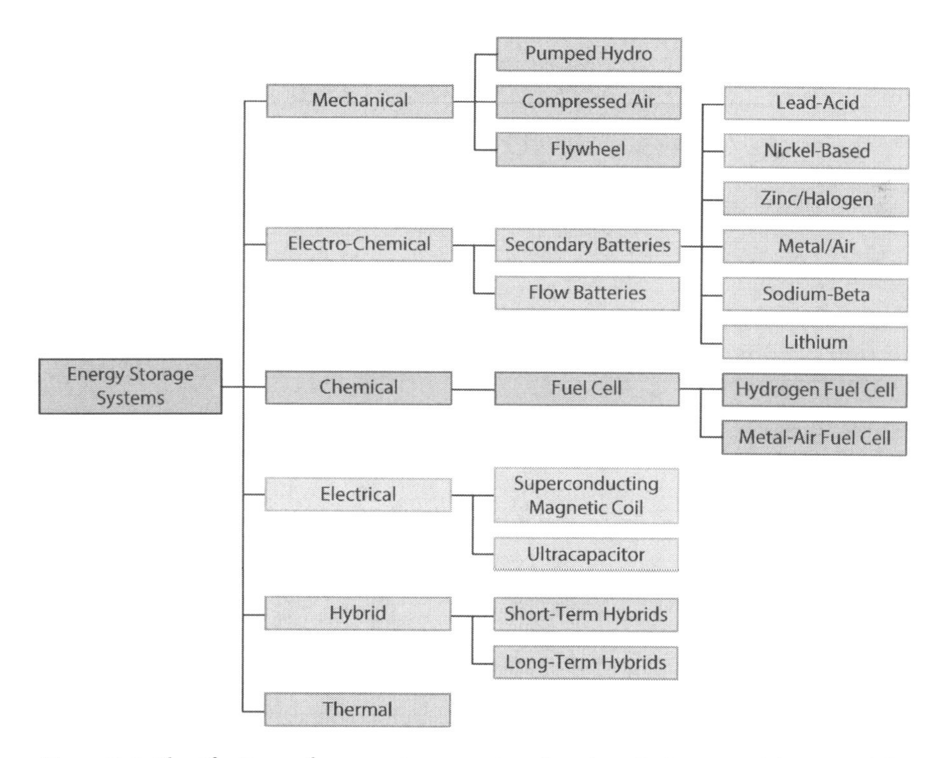

Figure 3.4 Classifications of energy storage systems based on their composition material.

3.3 Modeling of ESSs Applied in EVs

3.3.1 Mechanical Energy Storages

The mechanical storage systems can be classified into the flywheel, compressed air, and pumped hydro storage. The most common mechanical storage system is pumped hydro storage, which is applied for pumped hydroelectric power generations. In compressed air energy storage, compressed air is combined with natural gas, expanded, and later transformed into modified gas to supply a gas turbine to generate electricity. Besides, this kind of storage is appropriate for huge capacity electricity generation [17, 25]. The flywheel energy storage is described below.

3.3.1.1 Flywheel Energy Storages

Flywheel energy storages are proper for the EV application as well as power networks due to development in power electronic devices. The energy efficiency and nominal power of the flywheel energy storage can be within the range of up to 95% and 45 MW, respectively. The flywheel consists of a rotating shaft in a chamber, conjugated bearings, and an energy transmission system. The energy retained by the continuously rotating flywheel is transformed into electrical energy. The energy generated from the flywheel is the rotational kinetic energy, which can be formulated by Eq. (3.1).

$$E = \frac{1}{2} J \times \omega^2 \quad and \quad J = \frac{1}{2} m \times r^2 \tag{3.1}$$

where E represents the kinetic energy, J represents the inertia moment, ω, m, and r represent the angular velocity, mass, and radius of the flywheel's shaft, respectively.

It can be seen clearly from Eq. (3.1) that the generated power can be raised by enhancing the inertia moment or angular velocity of the flywheel's shaft. The major advantages of the flywheel energy storage are large power and energy density, no depth of discharge (DOD) impact, less operating cost, and long life. However, the flywheel energy storage has huge self-discharging characteristics due to bearing friction losses [26].

3.3.2 Electrochemical Energy Storages

All regular rechargeable batteries are using electrochemical storage systems (ECSSs). Especially, the flow batteries (FBs) and the secondary batteries can be identified as ECSS. In ECSSs, energy is converted from electrical form to chemical form of energy and vice versa via a reversible procedure with energy efficiency and small physical alternations. However, the chemical reaction may decrease cell lifetime and available energy. These kinds of batteries have the dual function of saving and releasing electrical power by modifying the charge and discharge phases with zero-emission and low maintenance [17, 27, 28].

3.3.2.1 Flow Batteries

The FBs are generally rechargeable and the potential power is restored in electroactive species. The electroactive species are dispersed in the liquid electrolyte in storage, then the liquid is pumped through an electrochemical cell to transform chemical energy into electric form. Redox flow batteries (RFB) and hybrid flow batteries (HFB) are illustrations of the FBs. The total scale of the storage of RFB determines the total feasible power of the battery. The RFBs represent a high life-cycle, high performance, resilience, and potential in power, which make RFBs desirable in any network. Illustrative models of RFBs are Fe-Cr, Fe-Ti, and poly S-Br batteries. The HFB cell can be identified as the concoction of a secondary battery (SB) and an RFB. In the RFB, the feasible capacity is determined by the physical scale of the electrochemical cell. The HFBs contain Zn-Br and Zn-Ce batteries. It should be noted that the FBs have a lifetime of 10–20 years, 5–15 h discharge range, and 60–70% efficiency rate [29]. Recently, the RFBs and the HFBs are being proposed for huge-scale power storage for local energy storage and utility-scale implementation for increasing power quality, emergency application, peak-load shaving, enhancing the security of power generation, and integration with RESs [30, 31].

3.3.2.2 Secondary Batteries

The SBs control the market for movable energy storage for EVs and other electric implementations. These batteries save required electricity taking the form of chemical energy and generate electricity via an electrochemical reaction procedure. The SB mainly includes two basic electrodes, namely, anode and cathode, electrolyte, a separator, and a shield. The SB generally has favorable features, in particular, the high-specific energy, the high-power-density, smooth discharge characteristic, reduced resistance, minor memory impact, and variety of temperature performance. However, major batteries include toxic materials. Therefore, the environmental effect during battery discharging must be taken into account. In most

EV implementations, the high-energy-density with high-specific power of electricity storage can be ensured by SBs due to the special advances in battery technologies and appropriate costs. The diverse kinds of EVs majorly contain lead-acid (LA), nickel-based (namely, Ni-Cd, Ni-Fe, Ni-Zn, Ni-H_2, and Ni-MH), metal-air-based (namely, Al-Air, Fe-Air, and Zn-Air), zinc-halogen-based (namely, Zn-Br_2 and Zn-Cl_2), sodium-beta (namely, Na-S and Na-$NiCl_2$), ambient temperature lithium (namely, lithium-polymer (Li-poly), lithium-ion (Li-ion)), and high-temperature lithium (namely, Li- Al-FeS and Li-Al-FeS_2) batteries [32, 33].

- **Lead-Acid Batteries**. The LA batteries are normally applied in every internal combustion engine (ICE) vehicle as a starter and are commonly employed for emergency power generators and renewable energy-based storage due to their robustness, safe scheduling, temperature endurance, and reduced cost. This kind of battery contains Pb as an anode electrode, PbO_2 as a cathode electrode, and H_2SO_4 solution as an electrolyte. The electrochemical reaction occurs in the LA batter as indicated in Eq. (3.2).

$$Pb + 2PbO_2 + 2H_2SO_4 \Leftrightarrow 2PbSO_4 + 2H_2O \tag{3.2}$$

  This kind of battery operates for about 6–15 years with an optimum of 2,000 life cycles at 80% DOD and ensures 75–90% efficiency. The starting-lighting-ignition (SLI) and UPS batteries are the LA batteries with voltages of 6, 8, and 12 volts. In recent years, the valve-regulated LA (VRLA) has become a widespread battery for energizing the EVs due to its high-specific power, reduced initial cost, fast charge potential, and no necessity for periodical maintenance [34, 35].

- **Nickel-Based Batteries.** The nickel-based batteries apply nickel hydroxide as the cathode electrode and various anode electrode materials. Based on the proposed anode electrode materials, the nickel-based batteries can be categorized into Ni-Cd, Ni-Fe, Ni-Zn, Ni-H_2, and Ni-MH batteries. Basically, in nickel-based batteries, the active materials include the nickel oxyhydroxide as a cathode electrode, the potassium hydroxide solution as an electrolyte, and several metals such as Fe, Cd, Zn, MH, or H_2 as an anode electrode. The general electrochemical reaction that occurs in nickel-based batteries is demonstrated in Eq. (3.3).

$$\begin{gathered} X + 2NiO(OH) + 2H_2O \Leftrightarrow 2Ni(OH)_2 + X(OH)_2; \quad X = Fe / Cd / Zn \\ (M)H + 2NiO(OH) \Leftrightarrow M + Ni(OH)_2 \\ H_2 + NiO(OH) \Leftrightarrow Ni(OH)_2 \end{gathered} \tag{3.3}$$

Ni-Fe and Ni-Zn batteries are less favorable in EV implementations due to reduced specific power, higher cost, reduced life cycle, and the high necessity for periodical maintenance. Ni-Fe and Ni-Zn batteries ensure 75% power efficiency [36]. The Ni-Cd and Ni-MH are presently applied to energize EVs since they have a considerable number of life cycles (about 2,000) and power densities. However, Ni-Cd has a significant memory impact and price, which is 10 times more than the LA battery's price. On the other hand, the Ni-MH has reduced memory impact, minor environmental aspects, and a huge scheduling temperature range. Although Ni-MH releases heat during operation and requires a complicated algorithm and a high-cost charger, it is more appropriate than Ni-Cd for energizing EVs. The Ni-H_2 has a large potential rate, a long-life cycle, and endurance to overcharging/over-discharging without harm. However, this kind of battery is costly, has self-discharging behavior that is related to the H_2 pressure, and has reduced volumetric power density [37, 38].

- **Zinc-Halogen Batteries.** The zinc-halogen batteries contain Zn-Cl_2 and Zn-Br_2, which are applicable as the ESSs in EVs. The Zn-Cl_2 has a high-power density of 95 Wh/L and a reduced specific energy of 50 W/kg. Furthermore, the Zn-Cl_2 became suitable for power networks due to the growth alternations for controlling and periodical maintenance. The Zn-Br_2 batteries are applicable for the ESSs in EVs due to their high specific power of about 70 Wh/kg, swift charging potential, and reduced cost of applied material. However, this kind of battery has currently become sluggish in EV implementation due to their reduced specific power of 80 W/kg, much reactivity of bromine, and huge scale for electrolyte temperature regulation and circulation [39, 40]. The general electrochemical reaction in the Zi-Br_2 battery is presented in Eq. (3.4).

$$ZnBr_2(aq) \Leftrightarrow Zn^0 + Br(aq) \quad (3.4)$$

- **Metal-Air Batteries.** The metal-air batteries contain metal electrodes as an anode and oxygen from a boundless air supply as a cathode. Ca, Li, Mg, Al, Fe, and Zn are applied as anode metals in the metal-air batteries. Among these items, the lithium-air (Li-Air) battery is majorly appropriate for EV implementations due to its increased theoretical specific power of 11.14 kWh/kg, which is almost 100 times more than other batteries. However, this kind of battery has an enhanced fire risk that could follow from the mixture of air and humidity [41].

  The calcium-air (Ca-Air) battery has increased energy density; however, it is relatively costly. In the majority of cases, the Ca is applied as alloy material in electrodes of the battery to provide high performance. Magnesium-air (Mg-Air) battery with increased specific energy of almost 700 W h/kg is developed with an Mg alloy anode instead of absolute Mg and soluble O reactant available in seawater for undersea medium implementation. The electrochemical iron-air (Fe-air) battery has a reduced specific power of 50–70 Wh/kg and lower cost compared to the other metal-air batteries. This battery is a capable power source for EV implementations due to its reduced life cycle and no distortion of active materials or framework for long electrical cycling. The aluminum-air (Al-air) battery has enhanced specific energy, Ahr capacity, and terminal voltage. However, these features are decreased due to the water usage during discharging. Innovative Al-air batteries are developed with Al alloy to prevent corrosion and to operate at about 98% coulombic efficiency over a huge current density variety. This kind of battery is typically applied to power marine or underwater automotive. The aluminum-oxygen (Al-O_2) battery is majorly feasible in other kinds given that the Al-O_2 compound generates almost three times the power per kilogram of oxygen as a hydrogen fuel cell. The zinc-air (Zn-Air) battery has many advantages of FCs and regular cells and is electrically and mechanically rechargeable. The developed electrically rechargeable Zn-air battery applies a bi-functional air electrode for an enhanced life cycle, and the mechanically rechargeable Zn-Air battery is also developed so that the discharged anode could be substituted to prevent shape disfigurement. For the enhanced performance applications, the hybrid design can be established with Zn-Air of high-specific energy and LA of high-specific power in the form of Zn-Air LA hybrid battery storages [42, 43].

The metal-air batteries are desirable for rechargeable storage implementations due to their reduced material cost and high specific energy. The general electrochemical reaction in the metal-air battery is described in Eq. (3.5).

$$4Me + nO_2 + 2nH_2O \Leftrightarrow 4Me(OH)_n \tag{3.5}$$

where Me is metal, e.g., Ca, Li, Mg, Al, Fe, and Zn, and n represents the number that relies on the valence alternation for the metal oxidation.

- **Sodium-Beta Batteries.** Sodium (Na) is one of the favorite materials for the anode side of the battery. The sodium-beta batteries are unique batteries that employ solid electrolytes. These kinds of batteries apply beta-alumina (β-Al_2O_3) as the electrolyte, which presents excellent Na+ conductivity and electric isolation at increased temperatures. Based on the cathode materials, the sodium beta batteries can be categorized into sodium-sulfur (Na-S) and sodium metal halide batteries. This kind of battery has an increased temperature and manages in the range of 300–350°C. Moreover, it has sufficient power and energy density in the ranges of 140–230 Wh/kg and 140–240 W/kg, respectively, a long life cycle of about 4,000, and high-power efficiency of 75–95%, and also it is inexpensive and secure. The Na-S battery contains the molten solid sodium as anode and the molten sulfur as cathode and is separated by electrolytes of solid beta alumina ceramic [44]. The general electrochemical reaction in the Na-S battery is presented in Eq. (3.6).

$$Na + xS \Leftrightarrow Na_2S_x \tag{3.6}$$

where x can be in the range of [3–5].

The sodium-metal halide battery technologies have effectively energized the EVs since they have a greater cell voltage than the Na-S batteries. This kind of battery is popular as the Zero Emission Battery Research Activity (ZEBRA). A sodium-metal chloride (Na-$MeCl_2$) battery normally operates at a large temperature range of 200–350 °C. The ZEBRA batteries have the major desirable temperature for the EV application due to their high energy density, reduced corrosion, basic safety, and enhanced tolerance to over-charging and over-discharging than the Na-S due to the semi-solid cathode, extended life cycle, and decreased cost than other kinds of batteries. However, the ZEBRA batteries have a relatively reduced specific power (150 W-kg), and they require thermal control and self-discharge. The ZEBRA batteries such as Na-$NiCl_2$, Na-$FeCl_2$, and Na-Ni-$FeCl_2$ are acceptable for energy storage implementations. The major diversity between the two kinds of sodium-beta batteries is the further application of sodium aluminum tetrachloride ($NaAlCl_4$) as the secondary electrolyte in the ZEBRA batteries [45, 46].

- **Lithium Batteries.** Lithium batteries are favorable batteries for the ESSs of the EVs due to their enhanced energy density, enhanced specific power and energy, and reduced weight. Also, lithium batteries have no memory impacts and no harmful aspects, in contrast to lead or mercury. However, this kind of battery is more expensive than other kinds of batteries. It usually costs about $200 for 1400/kWh and requires protection systems for secure management and cell balancing systems to provide stable battery performance at a similar voltage and charge level.

  Lithium batteries are developed for both high and ambient temperatures. Other than sodium-beta batteries, lithium-aluminum-iron mono-sulfide (Li-Al-FeS) and lithium-aluminum-iron-disulfide (Li-Al-FeS_2) are defined as high-temperature lithium batteries. These lithium-sulfur batteries have enhanced power capacity and the decreased weight among all other lithium batteries. These kinds of batteries have a relatively reduced life cycle and require thermal control. The lithium-sulfur batteries normally work at a temperature range of 375–500 °C. The high-temperature lithium-sulfur batteries contain the Li-Al alloy as an anode, the iron sulfide as cathode, the molten lithium-chloride-potassium-chloride as an electrolyte, and a partitioner. In these kinds of batteries, the Li-Al alloy is applied to manage the lithium activity, and iron sulfide is applied to avoid corrosion of iron. The general electrochemical reactions in both kinds of lithium-sulfur batteries are described in Eqs. (3.7) and (3.8).

$$2Li - AlFeS \Leftrightarrow 2Al + Fe + Li_2S \tag{3.7}$$

$$2Li - Al + FeS_2 \Leftrightarrow 2Al + Fe + Li_2FeS_2 \tag{3.8}$$

The other lithium batteries that work commonly at ambient temperature are majorly Li-poly and Li-ion batteries, which are significantly developed for EV implementations. The Li-ion batteries are well known for ESSs and mobile electric devices due to their small size, reduced weight, and good capacity. The Li-ion battery has high power and energy density, decreased self-discharge, and extensive lifetimes. However, the life cycle of the Li-ion battery is influenced by temperature. Based on the cathode electrode, The Li-ion batteries can be categorized into lithium manganese oxide ($LiMn_2O_4$), lithium cobalt oxide ($LiCoO_2$), lithium iron phosphate ($LiFePO_4$), lithium nickel cobalt aluminum oxide ($LiNiCoAlO_2$), lithium nickel-manganese-cobalt oxide ($LiNiMnCoO_2$), and lithium titanate ($Li_4Ti_5O1_2$) batteries. The general electrochemical reaction in the Li-Ion battery is presented in Eq. (3.9).

$$LiMeO_2 + C \Leftrightarrow Li_{1-x}MeO_2 + Li_xC \tag{3.9}$$

Currently, the Li-ion battery technologies are being designed for the next generations of EVs. Accordingly, the high-power Li-ion technologies are being developed for the ESSs of the

EVs to provide the specific energy and power demand of the EVs [47–49].

- **Iron Oxide Batteries.** A group of transitional metal oxides has received increased attention as potential anode materials for novel Li-ion batteries. Among the Li-ion batteries, iron oxides, specially Fe_2O_3 and Fe_3O_4, have received excellent attention due to their enhanced theoretical capacity of almost ≈1000 mAh g^{-1}, high multitude, non-toxicity, high corrosion resistance, and decreased manufacturing cost. The enhanced safety related to the higher lithium insertion potential and the non-flammable manner is other advantages of iron-oxide-based anodes for bulk utilizations. The lithium storage system of iron oxides is based on a redox conversion reaction, where the iron oxides are decreased to metallic nanoclusters dispersed in a Li_2O matrix upon lithiation and are then reversibly restored to their initial oxidation states during delithiation. For instance, the reaction mechanism of Fe_2O_3 can be represented by Eq. (3.10).

$$Fe_2O_3 + 6Li^+ + 6e^- \Leftrightarrow 3Li_2O + 2Fe \tag{3.10}$$

The forward supplanting reaction of Fe_2O_3 with Li^+ is thermodynamically practicable, and the structure of Fe^0 from Fe^{3+} includes multiple-electron transfer per metal atom, resulting in an enhanced theoretical capacity of lithium storage. However, it seems that the reverse extraction of Li^+ from the Li_2O matrix is thermodynamically unfeasible. Furthermore, iron oxides endure from insignificant cyclability, which is relatively down to the drastic volume change during the charging and discharging procedures. The reduced conductivity of iron oxides also creates extra performance degradation, especially when charge/discharge at high current densities [50, 51].

3.3.3 Chemical Storage Systems

The chemical storage systems (CSSs) save and return saved energy via chemical reactions. The FC is a usual kind of chemical storage that transforms the chemical energy of the fuel to electrical power constantly. The major variation between an FC and a typical battery is the procedure they provide an energy source. In the FC, the fuel and oxidant are provided outsourced to supply electricity. The main property of the FC is the feasible capability to supply electricity subject to the provided active materials. It can be mentioned that the FCs normally provide fuel efficiency of range 45–80%. The FC has been well known as an energy source that can decrease the utilization of fossil fuels and CO_2 emissions. Hydrogen FCs (HFCs) are common and accessible technologies in the market. The HFCs apply an integration of hydrogen and oxygen to generate electricity. This integration could be regenerable and bilateral from electricity and water. The FCs can be categorized into various kinds according to the fuel and oxidant integration, electrolyte, operational temperature, and implementations. These kinds of FCs contain phosphoric acid FC (PAFC), alkaline FC (AFC), solid polymer fuel cell-proton exchange membrane FC (SPFC-PEMFC), solid oxide FC (SOFC), regenerable FC (RFC), molten carbonate FC (MCFC), and direct methanol FC (DMFC). The basic chemical reaction in FCs is described by Eq. (3.11).

$$2H_2 + O_2 \Leftrightarrow 2H_2O + Electricity \tag{3.11}$$

PEMFC, AFC, and DMFC control at ambient temperature and are utilized for low and medium scales of power storage. Also, the PAFC applies medium temperature. These kinds of FCs are applied in transportation implementation due to their enhanced operational efficiency, easy design, and reduced emission. The MCFC and SOFC control at high temperatures in the range of 650–1100 °C. These two FCs can be applied to utilize huge-scale power storage. The DMFC applied methanol as fuel since it is simple to save energy than hydrogen. The DMFC is a high-power-density FC; however, its performance is low. The SOFC has enhanced fuel efficiency and better reliability than the DMFC; however, it is costly and requires high ranges of temperature. The SOFCs are applied in EV implementations as a prospective secondary energy source due to their enhanced efficiency. The SPFC has desirable advantages for EV implementations since it has the highest energy density among all kinds of FCs.

Another kind of FC is metal-air FC (MAFC). The MAFCs have the high-energy-density and reduced cost. Although MAFCs have recharging issues, investigations on its development are continued since this kind of FC is a proper candidate for ESSs in EVs [17, 52–54].

3.3.4 Electrical Energy Storage Systems

The electrical energy storage systems (EESSs) vary from the ESSs due to their storage technologies. An EESS saves electrical energy promptly through the electric field by partitioning charges/magnetic fields by flux. The UCs and superconducting electromagnets are identified as EESSs [17].

3.3.4.1 Ultracapacitors

An ultracapacitor (UC) is identical to a usual capacitor based on framework and function. However, a UC can have a large energy potential with the amount of thousand farads, which is named supercapacitor. The specific power of UC can be approximately in the range of 800–2500 W/kg with energy efficiency near 95%. The UC has a longer-term lifetime of approximately 40 years among all kinds of ESSs. The UC is highly applicable in EV applications due to its high-energy storage. It is worth mentioning that the UC does not need maintenance, and it provides temperature insensitivity and a long-term scheduling period. For fast charging/discharging applications, the UCs are employed as the appropriate energy storage during electric braking and as the local energy source during swift acceleration required for mounting in the storage systems of the EVs [55].

The UCs can be categorized into three classes such as pseudo-capacitors, electric double-layer capacitors (EDLC), and hybrids capacitors. The EDLC has an increased power density than other kinds of capacitors; however, it exhibits low specific energy of approximately 4–8 Wh/kg, high self-discharge, and large costs. Accordingly, the UCs are commonly applied with batteries, FCs, or other kinds of ESSs in EV implementations enabling a proper integration of high-power, high-energy density, and growth of total lifetime is realized. The amount of energy restored in the capacitor is directly relative to its capacitance and squarely relative to the voltage between electrodes [56, 57]. Besides, the capacity directly enhances with the growth of the electrode surface area and the permittivity coefficient of dielectric materials, and also with the reduction of the interval between electrodes, as demonstrated by Eq. (3.12).

$$\begin{aligned} W_C &= \frac{1}{2} C \times V^2 = \frac{1}{2} Q \times V; \\ Q &= C \times V; \\ C &= \frac{\varepsilon \times A}{d} \end{aligned} \tag{3.12}$$

where W_C represents the electrostatic energy, C represents the capacitance rate of the UC, V represents the voltage between the electrodes, Q represents the amount of charge, represents the permittivity coefficient of dielectric materials, A represents the surface area of the electrode, and d represents the distance between electrodes.

3.3.4.2 Superconducting Magnetic

The superconducting magnetic energy storages (SMESs) save electrical energy with the help of a magnetic field. The SMESs have an enhanced power efficiency of approximately 95%, and an extended life cycle of 110000, and a fast response of milli-seconds. The common power rating of the SMESs is in kW to MW. The SMES is applied for UPS applications and power quality investigations. The SMES is established in EV implementations with batteries. The required power saved by the SMES is directly relying on the self-inductance of the coil and square of current flow, as presented in Eq. (3.13) [58, 59].

$$W_L = \frac{1}{2} L \times I^2 \tag{3.13}$$

where W_L represents the energy saved in the inductive coil, L represents the self-inductance, and I represents the current flow.

3.3.5 Thermal Storage Systems

The thermal storage systems save the heat energy in solitary storage from the electric/solar heater for further application for generating electricity or diverse heating functions. Thermal storage can be identified in different manners, such as latent, sensible, and thermo-chemical absorption heat storage. Latent heat storage applies organic (namely, paraffin) and inorganic (namely, salt hydrates) and phase change substances to provide heat transfer during the phase change of the storage. Generally, latent heat storage has a large energy density and effective heat exchange potential at invariable temperatures. Furthermore, sensible heat storages are prevalent technology and have normally solid (namely, ground or concrete) or liquid (namely, water or thermo-oil) of storage mechanism. In the above-mentioned systems, thermal storage relies on the temperature alternation in the storage mechanism, and the feasible capacity relies on the specific mass and heat of the storage mechanism, as presented in Eq. (3.14). In EV applications, the automated thermoelectric supply system, which transforms waste heat into useful power, can be possibly applied to optimize total energy efficiency and cost [17, 60, 61].

$$E_T = V.K.(T_2 - T_1) \tag{3.14}$$

where E_T represents the thermal energy saved in the volume of V, K represents the specific heat, and T_1 and T_2 represent the temperatures pre- and post-recharge, respectively.

3.3.6 Hybrid Storage Systems

The hybrid ESSs (HSS) integrate the generated power of two or more ESSs electronically with further technical aspects. In the HSS systems, the integration of an ESS with the high-power density and an ESS with high energy density, or an ESS with fast response and an ESS with slow response, or an ESS with the high cost and an ESS with low cost are developed by power electronic formations to supply optimal power for required demand. The HSSs can be further categorized into integrations of battery + battery, battery + ultracapacitor, FC + battery, battery + SMES, battery + flywheel, CAES + battery, FC + UC, FC + flywheel, and CAES + UC as the short- and long-term applications for the HSS systems. It is worth noting that Zn-Air + NiMH, FC + VRLA, Zn-Air + VRLA, FC + NiMH, FC + Li-ion, and Zn-Air + Li-ion are common investigations of high specific energy and high specific power integrations. Also, NiMH + UC, VRLA + UC, and Li-ion + UC are commonly low- and high-demand integrations. Besides, FC + UHSF, FC + UC, and CAES + UC are for long-term investigations. Moreover, the HSS can capture the high-frequency variation of UC and the low-frequency variation of batteries. The integration of FC and UC is the most desirable hybrids for EV investigations due to their low power generation capability [17].

3.3.7 Modeling Electrical Behavior

Battery modeling is necessary to improve the understanding of battery energy storage systems, whether for transportation or grid storage. A battery system generally consists of battery cells, the working principle of which is demonstrated in Figure 3.5. Based on the requirements for output voltage, power, and energy capacities for the hybrid EVs, a

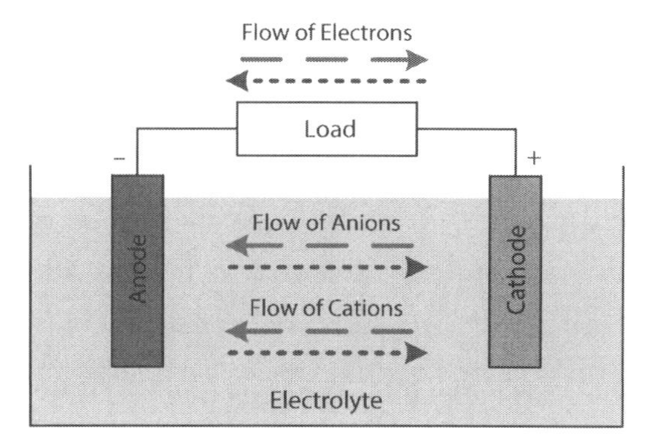

Figure 3.5 Scheme of an electrolysis system.

battery pack is configured by several cells connected in series or parallel, or both. The basic elements of a cell are anode, cathode, and separator. Upon the chemical reaction at the anode, electrons are released and flow to the cathode via an external circuit. Vital parameters for appointing the anode material can be presented as performance, high specific capacity, conductivity, stability, and reduced cost. The cathode is the electrode in which reduction (absorbing electrons) takes place. During discharge, the positive electrode of the cell is the cathode. During charge, the situation reverses and the negative electrode of the cell is the cathode. The cathode is appointed according to its voltage and chemical stability over time. Substances that release ions when dissolved in water are called electrolytes, which could contain a wide range of acids, bases, and salts, as they all give ions when dissolved in water. The electrolyte completes the cell circuit by transporting the ions. The electrolyte needs to be highly conductive, non-reactive (with the electrode materials), stable in properties at various temperatures, and cost-effective. A separator is a porous membrane inserted between electrodes of opposite charge [62, 63]. The key function of the separator is to keep the positive and negative electrodes discretely to prevent electrical short circuits and, simultaneously allow rapid transport of ionic charge carriers required to complete the circuit [64]. At a macroscopic level, a model has to be able to describe several key behaviors [7]:

- Voltage, which results from the contribution of several cells.
- Current, defined by external device or connection.

- Capacity, defined by the number of active materials in reagent form.
- State of charge, defined by the number of active materials in reagent form.
- Impedance, which represents the relation between the voltage and current.
- Losses, which represent the electrochemical efficiency in energy conversion.

The major issue with battery modeling is that several variables require to be estimated properly. From the itemized points above, only the voltage and the current can be calculated directly. Besides, these variables rely on each other and additional factors such as temperature and working period. An electrical circuit as demonstrated in Figure 3.6 can be applied to model the mathematical relations between the current and voltage determined at the battery terminals. In the presented voltage-current circuit, the voltage source defines the open-circuit voltage as a function of the SOC. The dynamics of a battery can be categorized as fast and slow elements. Generally, in the proposed model, the parameter *R-series* indicate the fast dynamic and the RC loop of the slow one. Furthermore, the losses can be also modeled by *R-series*. The parameters of the electrical circuit model are commonly recognized offline based on collected data from a test such as the hybrid pulse power characterization (HPPC) test. The HPPC test is designed to represent the charge and discharge power capabilities of an EV battery over a short period at various SOC levels under different operational temperatures [65].

Based on the equivalent electrical circuit, the mathematical relation between terminal voltage and current can be described as follows:

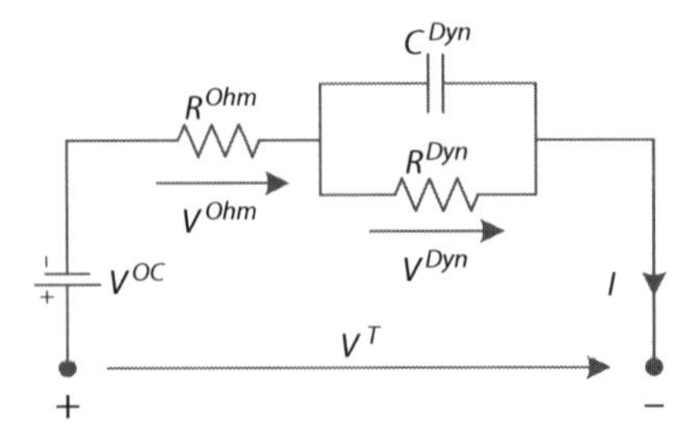

Figure 3.6 The equivalent model of a battery electrical circuit.

$$\text{Terminal Voltage:} \quad V^T = V^{OC} + V^{Ohm} + V^{Dyn} \tag{3.15}$$

$$\text{Open Circuit Voltage:} \quad V^{OC} = f(SoC, Tem) \tag{3.16}$$

$$\text{Voltage on Ohmic Resistor:} \quad V^{Ohm} = I.R^{Ohm}(SoC, Tem) \tag{3.17}$$

$$\text{Dynamic Voltage:} \quad I_R^{Dyn} = \frac{V^{Dyn}}{R^{Dyn}(SoC, Tem)}$$
$$I_C^{Dyn} = C^{Dyn}(SoC, Tem).\frac{dV^{Dyn}}{dt}$$
$$I = I_R^{Dyn} + I_C^{Dyn} \tag{3.18}$$

The dynamic voltage can be also formulated in a differential equation as follows:

$$\frac{dV^{Dyn}}{dt} + \frac{V^{Dyn}}{R^{Dyn}(SoC, Tem).C^{Dyn}(SoC, Tem)} = \frac{I}{C^{Dyn}(SoC, Tem)} \tag{3.19}$$

The final differential equation of the electrical circuit can be represented by Eq. (3.20).

$$\frac{dV^T}{dt} + \frac{V^T}{R^{Dyn}.C^{Dyn}} = R^{Ohm}\frac{dI}{dt} + \frac{R^{Dyn} + R^{Ohm}}{R^{Dyn}.C^{Dyn}}I + \frac{V^{OC}}{R^{Dyn}.C^{Dyn}} \tag{3.20}$$

Where V^T is the terminal voltage of the battery (in V), I is the terminal current of the battery (in A), Tem is the temperature of the battery, V^{OC}, R^{Ohm}, R^{Dyn} and C^{Dyn} are the parameters of the circuit model which are the functions of the Tem and SoC, and can be determined based on the special tests. V^{OC} is the open-circuit voltage which means that when the circuit is open the voltage source defines the open-circuit voltage. V^{Ohm} is the voltage on the ohmic resistor that depends on the current and amount of the parameter R-series that indicate the fast dynamic. Also, R-series represents the losses. V^{Dyn} is the voltage of the RC loop that indicates the slow dynamic of the battery.

3.3.8 Modeling Thermal Behavior

Storage devices with high energy density have the potential for high heat generation. For example, a single Li-ion polymer cell can endure an approximate temperature growth of 5 to 25 K at a 1C discharge rate, according to the cell form and the corresponding connections. Based on the ambient conditions, there may be a requirement to omit or add heat to the installed battery to maintain the optimal temperature range and distribution. Existing thermal management techniques include applying liquids, insulations, and phase-change materials. Essential tools in automotive pack design and thermal management are thermal and performance models. Such models require input parameters like system and scheduling data. Thermal modeling is an important issue for batteries to ensure safe operation, high efficiency, and a long lifetime. In this regard, the thermophysical attributes of the batteries are very important for developing reliable and accurate thermal models. Furthermore, the operational temperature is essential since operating a battery storage system (BSS) at a low ambient temperature influences its maximum capacity, while operating a BSS at a high ambient temperature influences its lifetime and provides safety risks. Safety risks can be increased in EV applications based on the physical size of the BSS needed to supply the energy demand. Accordingly, modeling the thermal behavior of these systems is an essential concern before designing an appropriate thermal management system that would schedule safely and extend the lifetime of the installed BSS [7].

Concerning battery thermal modeling, several numerical approaches have been developed to investigate the thermal behavior of the batteries, such as lumped thermal models and 3D thermal models. These numerical approaches have some technical profits. Generally, lumped models are uncomplicated and cost-effective in computation interval, thus, they can be suitably connected to the electrical models. However, the lumped models encounter difficulty with predicting the operating temperature of the batteries when there are intense temperature gradients, which usually occurs in battery modules and packs integrated with a cooling system. On the other hand, the 3D thermal models are a computationally time-consuming and costly approach. However, they are regularly applied to design battery thermal modeling systems because of the capability of predicting the spatial temperature distribution. Accordingly, depending on the scale level, accuracy, time, and system application, diverse thermal models are required for thermal modeling of the EV batteries, and obtaining the thermophysical parameters of the battery before this application is significantly important.

According to the electrical circuit model and the thermal dynamics of the battery, the temperature and the power generating heat can be calculated based on the following equations. The internal heat generation in the BSS is assigned to the temperature-current, overpotential heat generation, and SOC change in entropy heat generation. Heat generation is always positive and depends on the internal resistance within the battery, kinetic aspects, and mass transport related to electrochemical reactions. Moreover, the released heat generated by entropy alternation can be positive or negative based on the operating time and depends on the reversible electrochemical reactions within the battery. Besides, another source of internal heat generation is the heat generation from the enthalpy of mixing; however, it is normally neglected in many thermal models. In these models, the equations for thermal behavior are based on the heat generation equations and the boundary condition equations [66, 67].

$$\text{Heat Generation [W]:} \quad H^{Gen,d} = H^{R,Ohm} + H^{Dyn} + H^{React} \tag{3.21}$$

$$\text{Ohmic Resistance Heat [W]:} \quad H^{R,Ohm} = I.V^{Ohm} = I^2.R^{Ohm}(SoC,Tem) \tag{3.22}$$

$$\text{Dynamic Heat [W]:} \quad H^{Dyn} = |I|.|V^{Dyn}| = |I|.\left|\int_0^t \left[\frac{I(t)}{C^{Dyn}(SoC,Tem)} - \frac{V^{Dyn}(t)}{R^{Dyn}(SoC,Tem).C^{Dyn}(SoC,Tem)}\right]dt\right| \tag{3.23}$$

$$\text{Initial Condition [W]:} \quad V^{Dyn}\,|_{t=0} = 0 \tag{3.24}$$

$$\text{Reaction Heat [W]:} \quad H^{React} = \frac{\Delta S.I.\eta^{Bat}.T}{F} \tag{3.25}$$

$$\text{Dissipated Heat [W]:} \quad H^{Dissip,d} = \left(T^{Coolant} - T\right)h^{Bat} \tag{3.26}$$

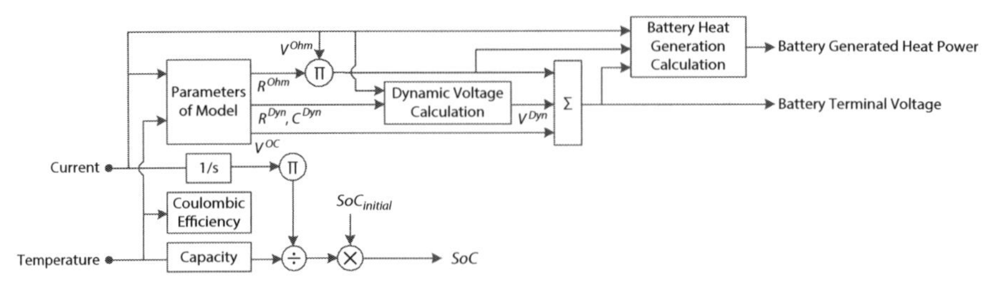

Figure 3.7 Electrical model of an EV battery [7].

$$\text{Battery Temperature [W]:} \quad T(t_i)+\int_{t_i}^{t}\frac{H^{Gen,d}+H^{Dissip,d}}{C^{ESS}.M^{ESS}}dt \tag{3.27}$$

Where $H^{Gen,d}$ is heat generation of the battery consists of $H^{R,Ohm}$, ohmic resistance heat, H^{Dyn}, dynamic heat which defines the generated heat from RC circuit that shows dynamic of battery, and H^{React} related to electrochemical reactions. ΔS is the delta entropy of reaction [in J/mol-K], η^{Bat} is the battery efficiency, and F is Faraday's constant equal to 96487 [C/mol], C^{ESS} is the specific heat capacity (J/K-kg) of the battery, $MESS$ is the mass (kg) of the battery, $T^{Coolant}$ is the temperature of the coolant in kelvin, and h^{Bat} is the heat transfer coefficient of the battery (W/K). According to the above-formulated equations, an electrical model of an EV battery can be applied by the diagram in Figure 3.7.

3.3.9 SOC Calculation

The SOC of the battery is applied to take into account the actual status of the battery capacity of an indicator. The SOC is identified as the ratio of the remained capacity to the nominal capacity of the battery. For the proper investigation of the EV scheduling and simulation at the system design level, the SOC calculation of battery can be realized as follows:

$$\text{SoC:} \quad SoC(t)=SoC(t_i)+\frac{1}{Cap_{A.hr}.3600}\int_{t_i}^{t}I(t).\eta^{Bat}.(SoC,Tem,sign[I(t)])dt \tag{3.28}$$

$$\text{Initial SoC:} \quad SoC(t_i)=SoC_i \tag{3.29}$$

Where $SoC(t_i)$ is the initial value of SOC at t equal to zero, $I(t)$ is the instant value of the charging/discharging current at time t, and $Cap_{A.hr}$ is the feasible capacity of the battery in ampere-hours. The SOC charge/discharge cycle is a relatively complex procedure; thus, the battery circuit model can quickly simplify the analysis of dynamic effects of the circuit. According to the circuit shown in Figure 3.6 and the Kirchhoff voltage and current methods, the circuit equations can be formulated by Eqs. (3.30) and (3.31).

$$V^T(t) = V^{OC}(SoC, Tem) - V^{Dyn}(t) - I.R^{Ohm}(SoC, Tem) \quad (3.30)$$

$$I(t) = \frac{V^{Dyn}}{R^{Dyn}} + C^{Dyn}.\frac{dV^{Dyn}}{dt} \quad (3.31)$$

Eq. (3.31) represents a first-order linear non-homogeneous equation. In this modeling, the following assumptions can be made: the battery current is negative when charging, the discharge current is positive, the initial voltage $V^T(0)$, time constant $\tau = R^{Dyn}.C^{Dyn}$. Accordingly, the following equations can be presented:

$$V^{Dyn}(t) = V^T(0).e^{-\left(\frac{t}{\tau}\right)} + I.R^{Dyn}.\left(1 - e^{-\left(\frac{t}{\tau}\right)}\right) \quad (3.32)$$

$$V^T(t) = V^{OC}(SoC, Tem) - I.R^{Ohm}(Soc, Tem) - \left[V^T(0).e^{-\left(\frac{t}{\tau}\right)} + I.R^{Dyn}(1 - e)^{-\left(\frac{t}{\tau}\right)}\right] \quad (3.33)$$

Under the condition of actual battery operation, it is difficult to accurately calculate the SOC of the power battery by applying a single linear mathematical model, since the battery is affected by non-linear factors such as high temperature, self-discharge, energy reduction, and discharge state [68]. To reduce the influence of the above nonlinear factors, a Kalman filter correction method can be applied to modify these factors. Kalman filter algorithm is mainly applied for real-time estimation of linear systems. Besides, non-linear systems require applying an extended Kalman filter algorithm for real-time estimation. According to the battery model equation and the definition of the SOC, the battery model of the discrete state space equation and its output equation can be designed [69].

$$\begin{bmatrix} SoC_{K+1} \\ V_{K+1}^{Dyn} \end{bmatrix} = \begin{bmatrix} 1 & 0 \\ 0 & e^{-\left(\frac{\Delta t}{\tau}\right)} \end{bmatrix} \begin{bmatrix} SoC_K \\ V_K^{Dyn} \end{bmatrix} \begin{bmatrix} -\left(\frac{\eta.\Delta t}{C}\right) \\ R^{Dyn}(1-e^{-\left(\frac{\Delta t}{\tau}\right)}) \end{bmatrix} I_k + Z_k \tag{3.34}$$

$$V_{sk}^T = V_{ck}^T(SoC_K, Tem) - I_K.R^{Ohm} - V_K^{Dyn} - S_k \tag{3.35}$$

In Eq. (3.34), Z_k is the systematic procedure noise, which is majorly caused by the measurement error of the system sensor. In Eq. (3.35), S_k is the system noise, which is majorly caused by the deviation between the current parameter and the target parameter.

Set the $X = \begin{bmatrix} x_1 & x_2 \end{bmatrix}^T$, $x_1(t) = SoC_{K+1}$, and $x_2(t) = V_{K+1}^{Dyn}$. By transforming Eqs. (3.34) and (3.35), Eqs. (3.36) and (3.37) can be defined.

$$\dot{x} = f(x,u) + Z_K \tag{3.36}$$

$$y = g(x,u) + S_K \tag{3.37}$$

By applying the Taylor formula to Eq. (3.36) and (3.37), Eqs. (3.38) and (3.39) can be achieved.

$$f(x,u) \approx f\left(x(t),u(t)\right) + \delta x \frac{\alpha f(x,u)}{\alpha x} | x(t),u(t) + \delta u \frac{\alpha f(x,u)}{\alpha x} | x(t),u(t) \tag{3.38}$$

$$g(x,u) \approx g\left(x(t),u(t)\right) + \delta x \frac{\alpha g(x,u)}{\alpha x} | x(t),u(t) + \delta u \frac{\alpha g(x,u)}{\alpha x} | x(t),u(t) \tag{3.39}$$

3.4 Characteristics of ESSs

The suitable selection of the installed ESSs for EV implementation majorly relies on their technical characteristics such as total generated power, energy capacity, discharge period, DOD, self-discharging manner, life cycle, total efficiency, and physical size. The aspects of total generated power in an ESS enclose the requirements of the conversion process and the requested demand in such a way that ESS can be discharged or recharged at its highest feasible amount. The energy capacity of an ESS can be identified as the total value of energy feasible in the installed system, which is available after a full charging of the ESS. The discharging power of ESS mainly relies on the network response and requested demand. The discharge period can be defined as the ratio of the power stored in an ESS to the highest feasible value of the power supplied by the system. The self-discharging manner is related to the value of the power lost over a period when the ESS is not in scheduling mode or is idle. The life cycle is also related to the sustainability of an ESS and mainly relies on the several occasions an ESS could supply power after every recharging action. The life cycle also relies on the utilized materials that compose the ESS. Total efficiency is related to the value of power that is supplied from the stored power in an ESS. Furthermore, the physical size of ESS is the essential factor of an ESS for EV implementations. The physical size determines the large power density of ESS in the reduced mass and volume of the proposed system. It can be mentioned that the performance of an ESS may vary from other ESSs based on their technical constraints. For cost-effective energy storage implementations in EVs, high power and energy density with a limited physical size are more important factors for ESSs. Moreover, minimal self-discharge, extensive cycle period, high performance, zero-emission, and lower maintenance are established in the industry, and choosing ESSs for EV energizing [6, 20].

3.5 Application of ESSs in EVs

Figure 3.8 demonstrates the scheduling time of diverse ESS technologies based on their energy delivery and the applications of various ESSs based on the general requirements of EVs and transportations. It is clear from Figure 3.8 that the scale of 10 kW to multiple 100 kW is demanded EV scheduling over a small period. The high-speed flywheels, secondary batteries, and UCs are usually employed in EV energizing. As illustrated in Figure 3.9, the allocation of ESS technologies could be investigated based on efficiency and the supposed life cycle. The flywheels, SMESs, and UCs have high performance and an extended life cycle at almost 80% DOD. It can be noted that Ni-Cd, Na-S, and Li-ion batteries have power efficiencies within the scale of 60–90% with a life cycle of 2,500–5,000. Also, it can be

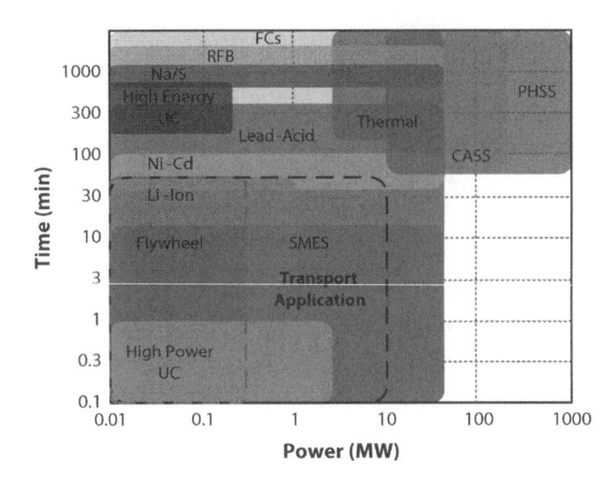

Figure 3.8 The ESS technologies [17].

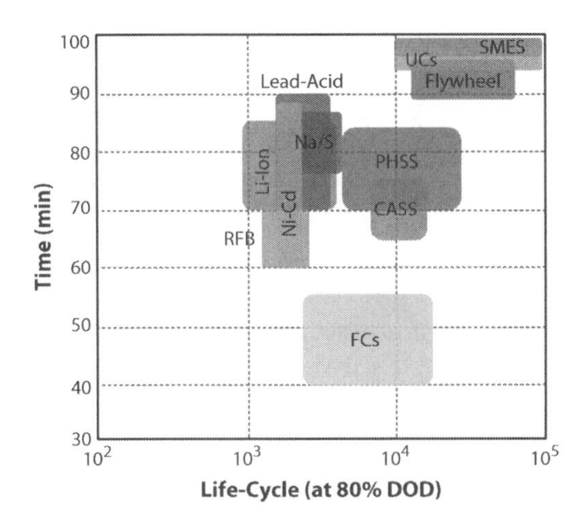

Figure 3.9 The ESS distribution based on the efficiency and life cycles [17].

seen that the presented FCs have low efficiency and high life cycles compared to other ESS technologies [17].

3.6 Methodologies of Calculating the SOC

The SOC of a battery is defined as the ratio of current capacity to rated capacity. It can be mentioned that rated capacity is provided by the producers and indicates the highest value of charge that can be saved in the battery system. In recent years, diverse approaches to calculating a battery SOC have been proposed. The majority of them rely on measuring some available parameters which vary with the SOC.

3.6.1 Current-Based SOC Calculation Approach

The current-based SOC calculation approach is also called the coulomb counting approach, which is the most common technique for SOC calculation. If an initial SoC_i is provided, the $SoC(t)$ can be formulated based on the current integral as follows:

$$SoC(t) = SoC_i + \frac{\eta_{bat}(I(t), Tem)}{Cap(Tem).3600} \int_{t_0}^{t} I(t)dt \tag{3.40}$$

where $Cap(Tem)$ and $\eta_{bat}(I(t), Tem)$ are the capacity and coulombic efficiency of the battery, which are functions of battery temperature, Tem. An implementation diagram for this approach is demonstrated in Figure 3.10. In this diagram, the capacity of the battery is found by a 1D look-up table, and the coulombic efficiency is indexed by a 2D look-up table. The current-based SOC calculation approach is comparably easy and reliable through the following three requirements that are met under scheduling conditions [70, 71]:

- The initial SOC point can be set precisely;
- The current going through the battery can be measured accurately;
- The capacity of the battery can be obtained correctly in real time.

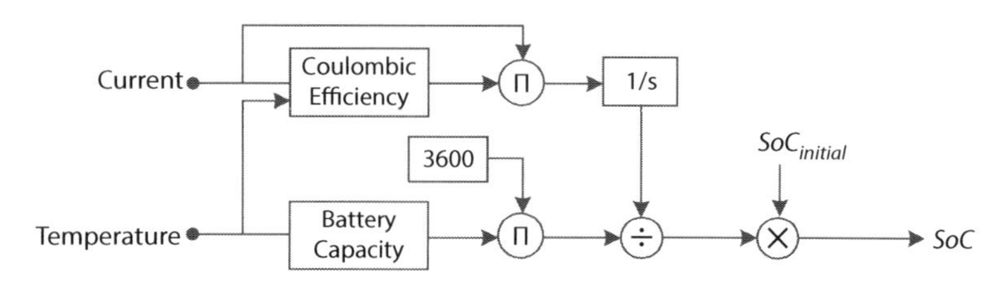

Figure 3.10 Diagram of current-based SoC calculation.

3.6.2 Voltage-Based SOC Calculation Approach

The voltage-based SOC calculation approach determines the SOC based on the measured or calculated open-circuit voltage of the battery. The open-circuit voltage, V^{OC}, can be defined as the terminal voltage of the battery when there is no connected external demand. The V^{OC} can be defined by the modified Nernst equation as follows:

$$V^{OC} = V^{O} - \frac{R_g Tem}{n_e F} \ln(Q); \qquad Q = f(SoC) \tag{3.41}$$

where F represents the Faraday constant (the number of coulombs per mole of electrons equal to $F = 9.6485309 \times 10^4\ Cmol^{-1}$, n_e represents the number of converted electrons, R_g represents the general gas constant equal to $R_g = 8.314472\ JK^{-1}mol^{-1}$, V^{O} represents the standard cell capacity, and Q represents the reaction quotient, which is a function of SoC.

Figure 3.11 demonstrated the relation between V^{OC} and the SoC of a Li-ion battery at a steady temperature. The main technical issue of the voltage-based SOC calculation approach is how to acquire V^{OC} accurately and robustly under functional conditions [72].

Based on the RC circuit model shown in Figure 3.6, the following equations can be presented at the provided SOC and temperature:

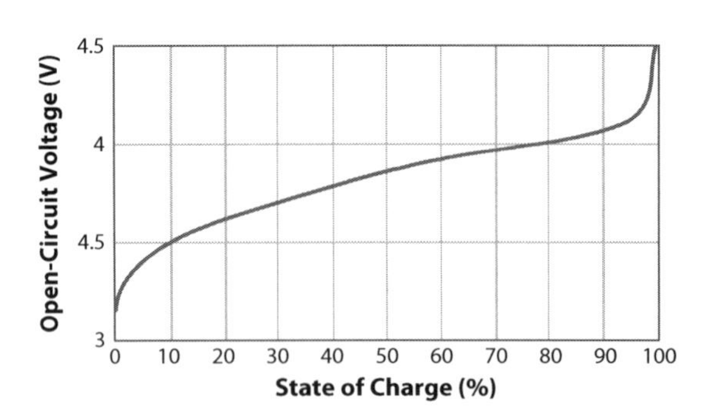

Figure 3.11 The relation between the open-circuit voltage and the state of charge of Li-ion battery.

$$V^T = V^{OC} + V^{Ohm} + V^{Dyn} = V^{OC} + R^{Ohm}.I + V^{Dyn} \tag{3.42}$$

$$C^{Dyn}\frac{dV^{Dyn}}{dt} + \frac{V^{Dyn}}{R^{Dyn}} = I \quad \rightarrow \quad \frac{dV^{Dyn}}{dt} + \frac{V^{Dyn}}{R^{Dyn}C^{Dyn}} = \frac{I}{C^{Dyn}} \tag{3.43}$$

Taking the derivative of Eq. (3.42) and assuming that V^{OC} is constant in a short time, Eq. (3.44) can be obtained.

$$\frac{dV^T}{dt} = \frac{dV^{OC}}{dt} + \frac{dV^{Ohm}}{dt} + \frac{dV^{Dyn}}{dt} = R^{Ohm}\frac{dI}{dt} + \frac{dV^{Dyn}}{dt} \tag{3.44}$$

Then, by dividing Eq. (3.42) by $R^{Dyn}C^{Dyn}$, Eq. (3.45) can be resulted.

$$\frac{V^T}{R^{Dyn}C^{Dyn}} = \frac{V^{OC}}{R^{Dyn}C^{Dyn}} + \frac{V^{Ohm}}{R^{Dyn}C^{Dyn}}.I + \frac{V^{Dyn}}{R^{Dyn}C^{Dyn}} \tag{3.45}$$

By adding Eq. (3.44) and (3.45), Eq. (3.46) can be obtained.

$$\frac{dV^T}{dt} + \frac{V^T}{R^{Dyn}C^{Dyn}} = R^{Ohm}\frac{dI}{dt} + \frac{dV^{Dyn}}{dt} + \frac{V^{OC}}{R^{Dyn}C^{Dyn}} + \frac{R^{Ohm}}{R^{Dyn}C^{Dyn}}.I + \frac{V^{Dyn}}{R^{Dyn}C^{Dyn}} \tag{3.46}$$

By substituting Eq. (3.43) to Eq. (3.46), Eq. (3.47) can be presented.

$$\begin{aligned}\frac{dV^T}{dt} + \frac{V^T}{R^{Dyn}C^{Dyn}} &= R^{Ohm}\frac{dI}{dt} + \frac{I}{C^{Dyn}} + \frac{V^{OC}}{R^{Dyn}C^{Dyn}} + \frac{R^{Ohm}}{R^{Dyn}C^{Dyn}}.I \\ &= R^{Ohm}\frac{dI}{dt} + \left(\frac{R^{Dyn} + R^{Ohm}}{R^{Dyn}C^{Dyn}}\right).I + \frac{V^{OC}}{R^{Dyn}C^{Dyn}}\end{aligned} \tag{3.47}$$

By discretizing Eq. (3.47), the difference equation can be described as follows:

$$\frac{V^T(k)-V^T(k-1)}{\Delta t}+\frac{V^T(k-1)}{R^{Dyn}C^{Dyn}}=R^{Ohm}.\frac{I(k)-I(k-1)}{\Delta t}+\left(\frac{R^{Dyn}+R^{Ohm}}{R^{Dyn}C^{Dyn}}\right).I(k-1)+\frac{V^{OC}}{R^{Dyn}C^{Dyn}}$$

$$\rightarrow \quad V^T(k)=(\quad).V^T(k-1)+R^{Ohm}.I(k)+\left(\frac{(R^{Dyn}+R^{Ohm})\Delta t}{R^{Dyn}C^{Dyn}}-R^{Ohm}\right).I(k-1)+\frac{\Delta t.V^{OC}}{R^{Dyn}C^{Dyn}} \tag{3.48}$$

where Δt represents the sampling time. This equation can be further formulated as follows:

$$V^T(k)=\alpha_1.V^T(k-1)+\alpha_2.I(k)+\alpha_3.I(k-1)+\alpha_4 \tag{3.49}$$

where

$$\alpha_1=\left(1-\frac{\Delta Tem}{R^{Dyn}C^{Dyn}}\right);\quad \alpha_2=R^{Ohm};\quad \alpha_3=\frac{(R^{Dyn}+R^{Ohm})\Delta Tem}{R^{Dyn}C^{Dyn}}-R^{Ohm};\quad \alpha_4=\frac{\Delta Tem.V^{OC}}{R^{Dyn}C^{Dyn}}$$

The parameters α_1, α_2, α_3, and α_4 can be determined in real time according to the measured battery terminal voltage and the current flow via a recursive calculation approach. Once these parameters are defined, the electrical circuit model parameters can be attained by solving equations as follows:

$$R^{Ohm}=\alpha_2;\quad V^{OC}=\frac{\alpha_4}{1-\alpha_1};\quad R^{Dyn}=\frac{\alpha_3+\alpha_1.\alpha_2}{1-\alpha_1};\quad C^{Dyn}=\frac{\Delta Tem}{\alpha_3+\alpha_1.\alpha_2} \tag{3.50}$$

3.6.3 Extended Kalman-Filter-Based SOC Calculation Approach

The Kalman filter provides a technique of calculating the states of a linear dynamic system, while an extended Kalman filter (EKF) can be applied to calculate the states of a nonlinear system. The EKF is a well-known algorithm largely used to estimate the state of a dynamic system characterized

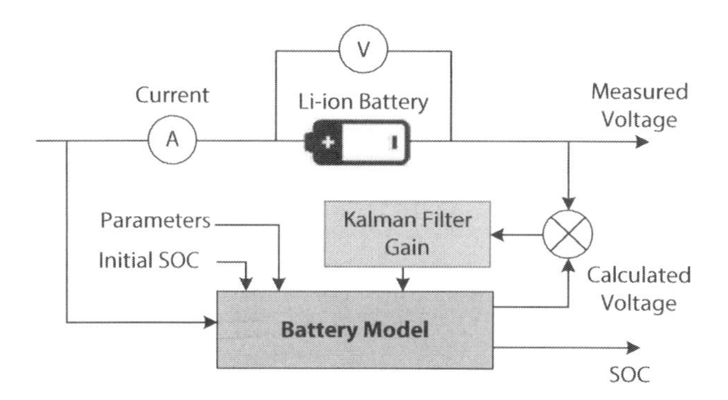

Figure 3.12 Block diagram of extended Kalman filter.

by noisy measurements. At the same time, this method may also be used to perform system parameters identification starting from experimental data [73]. As shown in Figure 3.12, the SOC estimation problem is addressed, to improve battery usage and vehicle power management. The EKF is considered regarding the electrical model in Figure 3.6.

Here it is proposed with the addition of the Gaussian noise on the state and the output equation, as Kalman assumption:

$$x(k+1) = Ax(k) + Bu(k) + v(k) \tag{3.51}$$

$$v(k) = g(x(k), u(k)) + w(k) \tag{3.52}$$

where $v(k) \sim N(0;Q)$ and $w(k) \sim N(0;R)$ are white Gaussian noises with zero means and covariance matrices Q for the model state equation and R for the output relation, respectively. The estimator equations can be summarized as follows:

Prediction Step:	$\check{x}^{-1}(k) = A\check{x}(k-1)A^T + Bu(k-1)$	(3.53)
	$P^{-1}(k) = AP(k-1)A^T + Q$	(3.54)

Correction Step:	$L(k) = P^{-1}(k)C(k)^T[C(k)P^{-1}(k)C(k)^T + R]^{-1}$	(3.55)
	$\breve{x}(k) = \breve{x}^{-1}(k) + L(k)\left[V^{\exp}(k) - g(\breve{x}^{-1}(k), u(k))\right]$	(3.56)
	$P(k) = (1 - L(k)C(k))P^{-1}(k)$	(3.57)

where $V^{\exp}(k)$ is the experimental voltage measurement. Matrices $L(k)$ and $P(k) = E[e(k) - e^T(k)]$ are respectively the Kalman gain and the covariance matrix of estimation error $e(k) = x(k) - \breve{x}(k)$. Matrix $C = \frac{\partial g(x)}{\partial x}$, used in the filter equations (in correction step) is the linearized output matrix.

Once the state-space equation is correctly established and the SOC is properly defined as a state, the EKF technique can be utilized to attain the SOC [7, 70]. If it is assumed that a battery is modeled by the two RC pair electrical circuit model shown in Figure 3.13, where one RC pair demonstrates the fast dynamic response behavior such as the charge transfer procedure and the other RC pair demonstrates the slow dynamic response behavior such as the solid diffusion procedure [74]. The detailed relationships between variables can be described as follows:

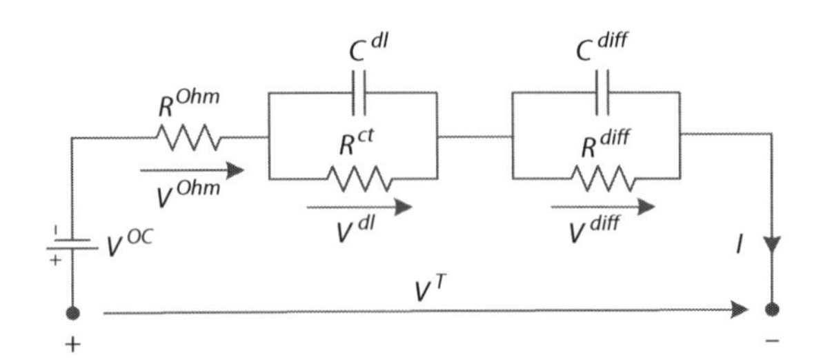

Figure 3.13 The electrical circuit equivalent model of the battery with two RC pair.

$$
\begin{aligned}
&SoC: && SoC(t)=\int_{t_i}^{t}\frac{\eta_{bat}\Delta t}{Cap}I.dt && \rightarrow && \frac{dSoC}{dt}=\frac{\eta_{bat}\Delta t}{Cap}I \\
&V^{diff}: && C^{diff}\frac{dV^{diff}}{dt}+\frac{V^{diff}}{R^{diff}}=I && \rightarrow && \frac{dV^{diff}}{dt}=-\frac{V^{diff}}{R^{diff}C^{diff}}=\frac{I}{C^{diff}} \\
&V^{dl}: && C^{dl}\frac{dV^{dl}}{dt}+\frac{V^{dl}}{R^{ct}}=I && \rightarrow && \frac{dV^{dl}}{dt}+\frac{V^{dl}}{R^{ct}C^{dl}}=\frac{I}{C^{dl}} \\
&V^{T}: && V^{T}=V^{OC}+V^{diff}+V^{dl}+V^{Ohm}=f(SoC)+V^{diff}+V^{dl}+R^{Ohm}.I
\end{aligned}
\tag{3.58}
$$

By discretizing Eq. (3.58) and defining the states as $x_1=SoC$, $x_2=V^{diff}$, and $x_3=V^{dl}$, the state-space equations can be calculated as follows:

$$
\begin{aligned}
&\text{State equation:} && \begin{bmatrix} x_1(k+1)\\ x_2(k+1)\\ x_3(k+1)\end{bmatrix}=\begin{bmatrix}1 & 0 & 0\\ 0 & 1-\frac{\Delta t}{R^{diff}C^{diff}} & 0\\ 0 & 0 & 1-\frac{\Delta t}{R^{ct}C^{dl}}\end{bmatrix}\begin{bmatrix} x_1(k)\\ x_2(k)\\ x_3(k)\end{bmatrix}+\begin{bmatrix}\frac{\eta_{bat}\Delta t}{Cap}\\ \frac{\Delta t}{C^{diff}}\\ \frac{\Delta t}{C^{dl}}\end{bmatrix}.I \\
&\text{Output equation:} && V^{T}(k)=f\left(x_1(k)\right)+x_2(k)+x_3(k)+R^{Ohm}.I(k)
\end{aligned}
\tag{3.59}
$$

where η_{bat} represents the coulombic efficiency, $V^{T}(k)$ and $I(k)$ represent the terminal voltage and current flow of the battery, respectively, C^{diff} represents the diffusion capacitance, R^{diff} represents the diffusion resistance, C^{dl} represents the double-layer capacitance, and R^{ct} represents the charge transfer resistance.

3.6.4 SOC Calculation Approach Based on the Transient Response Characteristics

This method approach calculates the SOC of the battery based on the specifications of transient time response. The battery model, shown in Figure 3.14, includes a potential with hysteresis, ohmic resistance, and a linear dynamic subsystem. The transient part of the time response is the part of the system response that equals zero as time increase. According to the linear system theory, the transient response characteristics of a system are

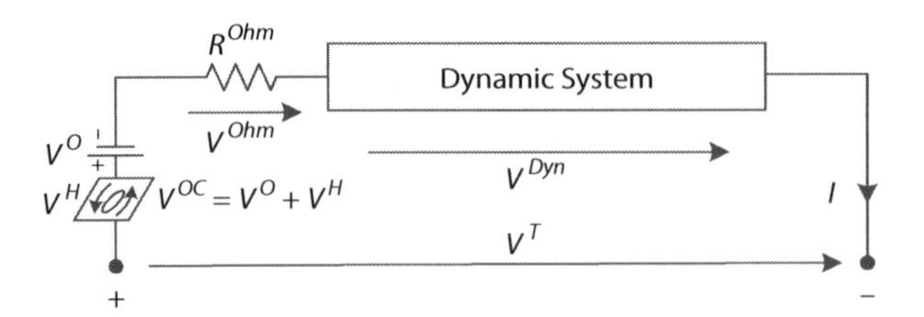

Figure 3.14 Overall electrical circuit equivalent model of the battery.

relevant to the locations of system poles. The closer the system pole is to the imaginary axis, the faster the system responds to the input [7, 75].

The methodology for calculating the SOC of a battery system based on transient response characteristics to a provided load current can be presented as follows:

1. Establish a linear equation describing the dynamics of the battery system, for instance, an appropriate order difference equation such as:

$$V(k) = a_1.V(k-1) + a_2(t).V(k-2) + ... + a_n(k-n).V(k-n) \\ +b_0.I(k) + b_1.I(k-1) + ... + b_m.I(k-m) \tag{3.60}$$

2. Estimate the parameters of the equation based on an online estimation algorithm such as a recursive least squares algorithm.
3. After obtaining the parameters, the system demonstrated in Eq. (3.60) can be introduced by a z-transfer function:

$$\frac{V(z)}{I(z)} = \frac{\breve{b}_1 z^{-1} + \breve{b}_2 z^{-2} + ... + \breve{b}_{m_0} z^{-m_0}}{1 + \breve{a}_1 z^{-1} + \breve{a}_2 z^{-2} + ... + \breve{a}_{n_0} z^{-n_0}} \tag{3.61}$$

where $V(z)$ is the voltage output of the system and $I(z)$ is the input current of the system.

4. The z-transfer function can be rewritten by Eq. (3.50) in pole/zero form:

$$\frac{V(z)}{I(z)} = \frac{k(z+z_1)(z+z_2)...(z+z_{m_0})}{(z+\rho_1)(z+\rho_2)...(z+\rho_{n_0})} \tag{3.62}$$

where z_i and ρ_j are the zero i and the pole j of the system.

5. Determine the dominant pole location of the battery system in the s-plane under the present operational conditions.
6. Calculate the SOC from the pre-established SOC-pole-location table based on the dominant pole location and the present operational temperature.

3.6.5 Fuzzy Logic

The fuzzy logic approach is an easy way to draw definite inferences from ambiguous or imprecise information. It resembles human decision-making with its capability to work from proximate data to find accurate solutions. Classical logic needs a deep understanding of a system, precise equations, and exact numeric values. Also, it provides complex systems to be modeled by the application of a higher level of abstraction originating from our knowledge and experience. Besides, it provides describing this knowledge with subjective concepts such as small, big, very, hot, bright red, a long time, fast or slow. This section briefly presents how to apply fuzzy logic approaches to estimate the SOC of a battery system [71]. The diagram of the fuzzy-logic-based battery SOC calculation algorithm can be defined as shown in Figure 3.13, in which battery dynamic behaviors are described by fuzzy logic. The SOC of a battery system under the present operational conditions can be determined by the following procedures.

1. Define the V^{Dyn} based on the measured $V^T(k)$, $I(k)$, $Tem(k)$, and the calculated $SoC(k-1)$ in the previous time step.
2. Subtract the measured $V^T(k)$ from the calculated V^{Dyn} applying fuzzy logic; that is, $V(k) = V^T(k) - V^{Dyn}$.
3. Construct the regression equation $V(k) = R^{Ohm}.I(k) - V^{OC}$.
4. Estimate V^{OC} and R^{Ohm} via a recursive least square estimation method.

5. Use a predetermined look-up table to find out $SoC(k)$ from the estimated V^{OC}.

3.6.6 Neural Networks

The advantages of neural networks significantly depend on their potential to indicate both linear and non-linear functions and their ability to learn functions directly from the data being modeled. The neural network approaches are applicable in calculating battery efficiency which highly relies on quantifying the result of many technical parameters, most of which cannot be realized with mathematical accuracy. The algorithms can be refined with the aid of experience attained from the performance of similar batteries [71].

3.7 Estimation of Battery Power Availability

In EV implementations, the feasible charge/discharge power can be presented as follows:

$$P_{\max}^{char}(t) = \min\left\{I_{V_{\max}}(t)V_{\max}, I_{\max\_char}V_{I_{\max\_char}}(t)\right\}$$
$$P_{\max}^{dischar}(t) = \min\left\{I_{V_{\min}}(t)V_{\min}, I_{\max\_dischar}V_{I_{\max\_dischar}}(t)\right\} \quad (3.63)$$

where $P_{\max}^{char}(t)$ is the maximum feasible charge power of the battery, $V_{\max}$ is the maximum acceptable terminal voltage, $I_{V_{\max}}(t)$ is the current corresponding to the terminal voltage on the maximum acceptable terminal voltage, $I_{\max\_char}$ is the maximum acceptable charge current, $V_{I_{\max\_char}}$ is the terminal voltage corresponding to the current on the maximum acceptable charge current, $P_{\max}^{dischar}$ is the feasible maximum discharge power of the battery, $V_{\min}$ is the minimum acceptable terminal voltage, $I_{V_{\min}}$ is the current corresponding to the minimum acceptable terminal voltage, $I_{\max\_dischar}$ is the maximum acceptable discharge current, and $V_{I_{\max\_dischar}}$ is the terminal voltage corresponding to the current on the maximum acceptable discharge current.

3.7.1 PNGV HPPC Power Availability Estimation Approach

The Partnership for a Novel Generation of Vehicles (PNGV) provided a power availability estimation approach based on the hybrid pulse power characterization (HPPC) test [65, 76]. In the PNGV HPPC approach, the battery terminal voltage can be defined by Eq. (3.64).

$$V^T(t+\Delta t) = V^{OC}\left(SoC(t+\Delta t)\right) + R^{\text{int}}.I(t+\Delta t) \tag{3.64}$$

where R^{int} is the internal resistance which is set to be constant at a given SOC and temperature with diverse charge/discharge conditions. From Eq. (3.64), the maximum charge/discharge current is formulated as follows:

$$\begin{aligned} I_{\max\_char}(\Delta t) &= \frac{V_{\max} - V^{OC}(SoC(t))}{R_{char}^{\text{int}}} \\ I_{\max\_dischar}(\Delta t) &= \frac{V^{OC}(SoC(t)) - V_{\min}}{R_{dischar}^{\text{int}}} \end{aligned} \tag{3.65}$$

Then, the maximum feasible charge/discharge power is calculated as follows:

$$\begin{aligned} P_{\max\_char}(\Delta t) &= V_{\max}.I_{\max\_char}(\Delta t) \\ P_{\max\_dischar}(\Delta t) &= V_{\min}.I_{\max\_dischar}(\Delta t) \end{aligned} \tag{3.66}$$

If it is considered that both operational voltage and current limits, Eq. (3.66) requires to be rewritten as follows:

$$\begin{aligned} P_{\max\_char}(\Delta t) &= V_{\max}.\min\left\{I_{\max\_char}(\Delta t), I_{\max\_char\_limit}(\Delta t)\right\} \\ P_{\max\_dischar}(\Delta t) &= V_{\min}.\min\left\{I_{\max\_dischar}(\Delta t), I_{\max\_dischar\_limit}(\Delta t)\right\} \end{aligned} \tag{3.67}$$

where $I_{\max\_char\_limit}(\Delta t)$ and $I_{\max\_dischar\_limit}(\Delta t)$ are the maximum feasible acceptable charge/discharge currents in the time interval Δt.

3.7.2 Revised PNGV HPPC Power Availability Estimation Approach

The PNGV HPPC approach considers that V^{OC} and R^{int} are constant in the time interval Δt, and if R^{int} is the internal resistance of a battery and Δt is very small, Eq. (3.67) can predict the maximum instant charge/discharge power. However, if Δt is very large, the consideration will not continue as the V^{OC} changes with time. To resolve this issue, a developed approach was proposed [65, 76]. By applying Taylor series expansion, the open-circuit voltage, V^{OC}, can be described as follows:

$$V^{OC}\left(SoC(t+\Delta t)\right)=V^{OC}\left(SoC(t)+\frac{\Delta t\times\eta_i}{3600\times Cap}I(t)\right)\approx V^{OC}\left(SoC(t)\right)+\frac{\partial V^{OC}}{\partial SoC}\frac{\Delta t\times\eta_i}{3600\times Cap}I(t) \tag{3.68}$$

Then, Eqs. (3.64) and (3.65) can be rewritten as follows:

$$V^{T}\left(t+\Delta t\right)\approx V^{OC}\left(SoC(t)\right)+\frac{\partial V^{OC}}{\partial SoC}\frac{\Delta t\times\eta_{bat}}{3600\times Cap}I(t)+R^{\text{int}}.I\left(t+\Delta t\right) \tag{3.69}$$

$$\begin{aligned}
I_{\max_char}(\Delta t)&=\frac{V_{\max}-V^{OC}(SoC(t))}{\dfrac{\partial V^{OC}}{\partial SoC}\dfrac{\Delta t\times\eta_{bat}}{3600\times Cap}+R_{char}^{\text{int}}}\\
I_{\max_dischar}(\Delta t)&=\frac{V^{OC}(SoC(t))-V_{\min}}{\dfrac{\partial V^{OC}}{\partial SoC}\dfrac{\Delta t\times\eta_{bat}}{3600\times Cap}+R_{dischar}^{\text{int}}}
\end{aligned} \tag{3.70}$$

The maximum feasible charge/discharge power in the time interval Δt can be calculated Eqs. (3.71) and (3.72).

$$\begin{aligned}
P_{\max_char}(\Delta t)&=V_{\max}.I_{\max_char}(\Delta t)\\
&=V_{\max}.\min\left\{\frac{V_{\max}-V^{OC}(SoC(t))}{\dfrac{\partial V^{OC}}{\partial SoC}\dfrac{\Delta t\times\eta_i}{3600\times Cap}+R_{char}^{\text{int}}},I_{\max_char_limit}(\Delta t)\right\}
\end{aligned} \tag{3.71}$$

$$P_{\max\_dischar}(\Delta t) = V_{\min}.I_{\max\_dischar}(\Delta t)$$
$$= V_{\min}.\min\left\{\frac{V^{OC}(SoC(t)) - V_{\min}}{\frac{\partial V^{OC}}{\partial SoC}\frac{\Delta t \times \eta_i}{3600 \times Cap} + R^{\text{int}}_{dischar}}, I_{\max\_dischar\_limit}(\Delta t)\right\}$$

(3.72)

It is worth noting that all input parameters in the above power calculation equations are functions of battery temperature as well.

3.7.3 Power Availability Estimation Based on the Electrical Circuit Equivalent Model

As the electrical circuit equivalent model shown in Figure 3.6 can describe the terminal behavior of the battery, the power availabilities of the battery can also be calculated from this model [65, 76]. To obtain the power availability calculation formula, the terminal voltage equation of the one RC pair electrical circuit equivalent model can be modified as follows:

$$\frac{dV^T}{dt} + \frac{V^T}{R^{Dyn}C^{Dyn}} = R^{Ohm}.\frac{dI}{dt} + \left(\frac{R^{Dyn} + R^{Ohm}}{R^{Dyn}C^{Dyn}}\right).I + \frac{V^{OC}}{R^{Dyn}C^{Dyn}}$$

(3.73)

If the maximum acceptable constant current, $I_{\max\_limit}$, is applied to the battery, the terminal voltage response can be described as follows:

$$\frac{dV^T}{dt} + \frac{V^T}{R^{Dyn}C^{Dyn}} = \left(\frac{R^{Dyn} + R^{Ohm}}{R^{Dyn}C^{Dyn}}\right).I_{\max\_limit} + \frac{V^{OC}}{R^{Dyn}C^{Dyn}} \quad (3.74)$$

The solution of Eq. (3.74) is:

$$V^T(t) = R^{Dyn}.I_{\max\_limit}\left(1 - e^{-\left(\frac{t}{R^{Dyn}C^{Dyn}}\right)}\right) + R^{Ohm}.I_{\max\_limit} = \left(\frac{R^{Dyn} + R^{Ohm}}{R^{Dyn}C^{Dyn}}\right).I_{\max\_limit} + V^{OC}$$

(3.75)

Accordingly, the maximum constant current-based charge/discharge power availabilities in the time interval Δt can be described as follows:

$$P_{\max\_char\_IB}(\Delta t) = R^{Dyn}.\left(I_{\max\_char\_limit}\right)^2.\left(1-e^{-\left(\frac{\Delta t}{R^{Dyn}C^{Dyn}}\right)}\right) + R^{Ohm}.\left(I_{\max\_char\_limit}\right)^2 + V^{OC}.I_{\max\_char\_limit} \quad (3.76)$$

$$P_{\max\_dischar\_IB}(\Delta t) = V^{OC}.I_{\max\_dischar\_limit} - R^{Ohm}.\left(I_{\max\_dischar\_limit}\right)^2 - R^{Dyn}.\left(I_{\max\_dischar\_limit}\right)^2.\left(1-e^{-\left(\frac{\Delta t}{R^{Dyn}C^{Dyn}}\right)}\right) \quad (3.77)$$

If the maximum acceptable constant voltage, $V_{\max\_limit}$, is applied to the battery, the terminal current response can be described as follows:

$$\frac{dI}{dt} + \left(\frac{R^{Dyn}+R^{Ohm}}{R^{Dyn}C^{Dyn}}\right).I + \frac{\left(V_{\max\_limit}-V^{OC}\right)}{R^{Ohm}R^{Dyn}C^{Dyn}} \quad (3.78)$$

Then, the solution of Eq. (3.78) is:

$$I(t) = \frac{V_{\max\_limit}-V^{OC}}{R^{Dyn}+R^{Ohm}} + \frac{\left(V_{\max\_limit}-V^{OC}\right)R^{Dyn}}{R^{Ohm}\left(R^{Dyn}+R^{Ohm}\right)}.e^{-\left(\frac{R^{Dyn}+R^{Ohm}}{R^{Ohm}R^{Dyn}C^{Dyn}}.t\right)} 60^{\circ} \quad (3.79)$$

Accordingly, the maximum constant voltage-based charge/discharge power availabilities in the time interval Δt can be described as follows:

$$P_{\max\_char\_VB}(\Delta t) = V_{\max\_char\_limit}.I + \frac{V_{\max\_char\_limit}-V^{OC}}{R^{Ohm}+R^{Dyn}}.V_{\max\_char\_limit} + \frac{\left(V_{\max\_char\_limit}-V^{OC}\right)R^{Dyn}}{R^{Ohm}\left(R^{Dyn}+R^{Ohm}\right)}.V_{\max\_char\_limit}.e^{-\left(\frac{R^{Dyn}+R^{Ohm}}{R^{Ohm}R^{Dyn}C^{Dyn}}\Delta t\right)} \quad (3.80)$$

$$P_{\max\_dischar\_VB}(\Delta t) = V_{\max\_dischar\_limit}.I + \frac{V_{\max\_dischar\_limit} - V^{OC}}{R^{Ohm} + R^{Dyn}}.V_{\max\_dischar\_limit} + \frac{\left(V_{\max\_dischar\_limit} - V^{OC}\right)R^{Dyn}}{R^{Ohm}\left(R^{Dyn} + R^{Ohm}\right)}.V_{\max\_dischar\_limit}.e^{-\left(\frac{R^{Dyn}+R^{Ohm}}{R^{Ohm}R^{Dyn}C^{Dyn}}\Delta t\right)} \tag{3.81}$$

The maximal charge/discharge power availabilities of a battery in the time interval Δt should be the minimum possible of the maximum constant current-based charge/discharge power availabilities and the maximum constant voltage-based charge/discharge power availabilities, that can be presented as follows:

$$\begin{aligned} P_{\max\_char}(\Delta t) &= \min\left\{P_{\max\_char\_IB}(\Delta t), P_{\max\_char\_VB}(\Delta t)\right\} \\ P_{\max\_dischar}(\Delta t) &= \min\left\{P_{\max\_dischar\_IB}(\Delta t), P_{\max\_dischar\_VB}(\Delta t)\right\} \end{aligned} \tag{3.82}$$

3.8 Life Prediction of Battery

Constructively monitoring the state of health (SOH) of the battery system significantly enhances the reliability of an EV [77]. Accurate life prediction of batteries is important to help assess battery quality in advance, improve long-term battery planning, and subsequently guarantee the safety and reliability of battery operations. In recent years, SOH estimation and remaining useful lifetime (RUL) prediction are two vital research aspects in the battery management systems. SOH is an indicator reflecting the health state of the battery in the short term while RUL is a long-term indicator that shows the remaining cycle life before SOH drops to a predefined threshold [78]. In this section, the life prediction of the battery is comprehensively reviewed.

3.8.1 Aspects of Battery Life

When a battery is employed under various conditions, the composition and decomposition reactions of the active material happen all the time. Generally, battery aging is majorly ascribable to the following three reasons:

1. Degradation of active material reactions with the electrolyte at electrode interfaces;
2. Self-degradation of the active material structure on cycling;
3. Aging of non-active components.

The main factors of a battery that have significant results on battery life are as follows:

3.8.1.1 *Temperature*

The battery operates in a hotter situation, the chemical reactions are faster, and the battery performance will be better observably. In contrast, unwanted reactions are also active when the temperature is higher. Shown in Figure 3.15 is a lead-acid battery capacity alter curve acquired through the year-round operation at diverse temperatures. It can be seen from Figure 3.15 that the battery has an enhanced capacity performance at 35°C than the lower temperature, however, the high-temperature operation is also attended by a short life and huge economic waste.

3.8.1.2 *Depth of Discharge*

When the other operating conditions are assessed, the transformation of active chemicals is proportionate to the DOD in the procedure of charging/discharging. Figure 3.16 represents how the battery cycle life changes with the DOD of a lead-acid battery. Notably, with the higher DOD at which the battery cycles, the battery cycle life subsides clearly [77].

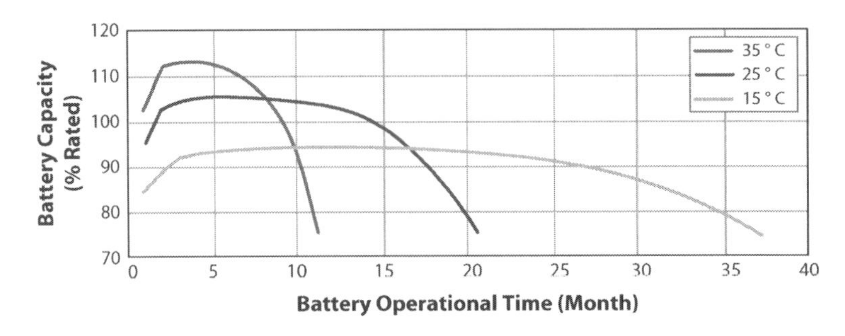

Figure 3.15 Battery capacity degradation at diverse temperatures [77].

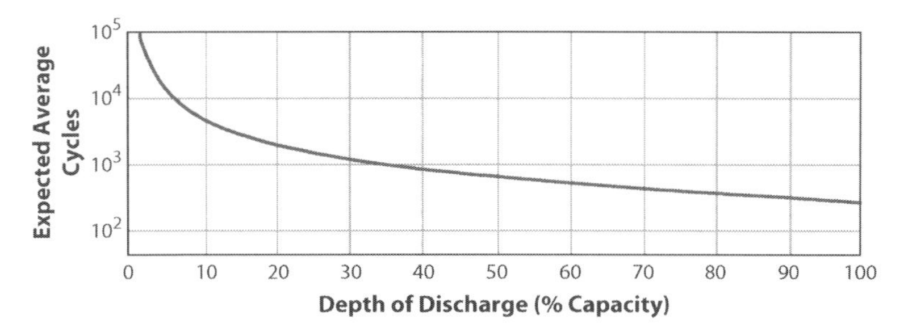

Figure 3.16 Cycle life vs. DOD curve for a LA battery [77].

3.8.1.3 Charging/Discharging Rate

For the lithium battery, how much lithium ions the battery anode material can carry per unit of time has an upper limit. Squeezing excessive current into the battery through the charging procedure will cause surplus lithium ions to deposit on the surface of the electrode to form the lithium metal layer which is called lithium plating. This undesirable chemical phenomenon is pursued by serious capacity loss and internal impedance growth. On the other hand, too high a discharge rate will mean that the chemical transformation procedure of the active material cannot meet the output of the battery current. One of the results is additional harmful chemical reactions that make further causing electrode crystal morphological change [77]. Figure 3.17 demonstrates the changing trend of battery capacity at different charge and discharge rates.

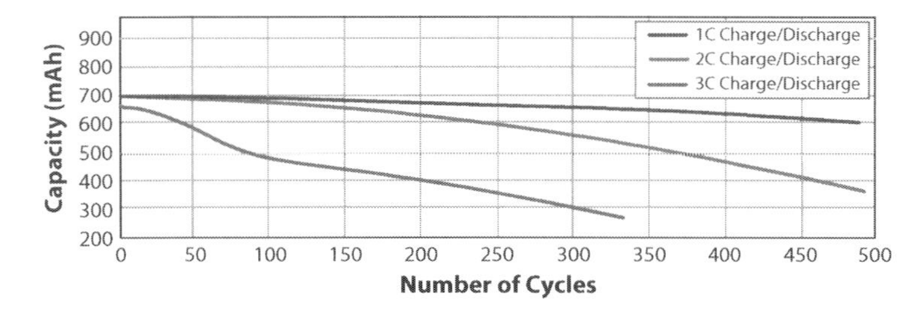

Figure 3.17 Battery capacity vs number of cycles curve under diverse rates [77].

3.8.2 Battery Life Prediction Approaches

There are three often-employed prediction approaches, namely, physic-chemical aging approach, event-oriented aging approach, and life determination based on the state of life (SOL). Various methods, including mechanism methods, the method based on SOL, and physic-chemical aging methods, have been developed to estimate and predict battery life conditions. These methods aim to capture the chemical and physical reactions in the battery, and the complexity inherent in the coupling of thermal and chemical heterogeneities within a cell [77].

3.8.2.1 Physic-Chemical Aging Method

In this model, the characterized deterioration index of battery performance gently enhances with the operating and cycling. When the failure threshold is reached, the battery achieves the end-of-life state and should be replaced. The degradation rate of the battery capacity related to the internal chemical reactions is impressed by many stress aspects incorporating charge/discharge rate, DOD, and operating temperature. Battery aging, resulting in a growth in the internal resistance, energy capability fading, and power capacity decay, emanates from several complicated mechanisms [77].

3.8.2.2 Event-Oriented Aging Method

In this method, a description of the particular event that causes the loss of life is defined. In general, each incident has a definiteness of the degree of damage. Monitoring the event status trend through the investigation of the equipment acquires the life attenuation of each event, and then gets the remaining life of the battery device. The events represented in the aging model are free of each other and the intensity of the recession caused by the same events at different life intervals is equivalent [77]. The aging analysis approach is appropriate to LA, Li-Ion, Na-S batteries, etc. The conversion coefficient $\alpha(x)$ is:

$$\alpha(x) = \frac{N_{BESS}(1)}{N_{BESS}(x)} \tag{3.83}$$

N_{BESS} represents the cycle life at the DOD x. Within the investigation interval, the number of converted charge/discharge cycles can be presented as follows:

$$N'_{BESS} = \sum_{1}^{n} \alpha(x_i) \tag{3.84}$$

Where n is the number of charge/discharge cycles, x_i is the DOD of every charge/discharge cycle. When $N'_{BESS} = N_{BESS}(1)$, the battery is in the end-of-life state. A unit degradation model is explicated to represent a detailed insight for the engineering investigations, as demonstrated in Eq. (3.85).

$$Loss_i = \frac{1}{N_{BESS}(x_i)} \tag{3.85}$$

Therefore, the battery life loss in the operation period can be assumed as the total of all the cycle degradation.

$$Loss = \sum_{1}^{n} Loss_i = \sum_{1}^{n} \frac{1}{N_{BESS}(x_i)} \tag{3.86}$$

The rain-flow calculating algorithm has been applied for separating the SOC curve to the full cycles and bringing convenience for succedent calculation. The benefits of the event-oriented aging model are that the computation is much simpler and it avoids boring measurements which are essential if need the battery internal performance. However, because the rain-flow calculating algorithm requires getting the whole running SOC curve to compute battery life loss, this method cannot be applied to battery life real-time monitoring. Also, because the model does not consider charge/discharge rate, the calendar life aging, and temperature results, the prediction accuracy cannot be guaranteed [77, 79].

3.8.2.3 Lifetime Prediction Method Based on SOL

3.8.2.3.1 Definition of State of Life

SOL is an appropriate characteristic for defining the degree of degradation of a battery in EV implementations. The SOL is a significant variable to warrant the good operation of electrically powered components and ensure that the battery operates in the best state. The SOL can be described in various ways, depending on specific properties and individual degradation. Besides, a battery may be feeble to fulfill one specification; however, it may still be ready to attain another. If the power availability deterioration is designed as the health index of a battery, the SOL can be defined as:

$$SOL_{P\_Based} = \frac{P_{ACT}(SoC,Tem) - P_{EOL}(SoC,Tem)}{P_{BOL}(SoC,Tem) - P_{EOL}(SoC,Tem)} \times 100\% \quad (3.87)$$

where $P_{BOL}(SoC,Tem)$ is the power availability at the beginning of life, $P_{EOL}(SoC,Tem)$ is the defined power availability at the end of life, and $P_{ACT}(SoC,Tem)$ is the actual power availability under the given (SoC,Tem) conditions.

If the internal impedance/resistance of a battery is developed as the health index of a battery, the SOL can be defined as follows:

$$SOL_{R\_Based} = \frac{R_{EOL}(SoC,Tem) - R_{ACT}(SoC,Tem)}{R_{EOL}(SoC,Tem) - R_{BOL}(SoC,Tem)} \times 100\% \quad (3.88)$$

where $R_{BOL}(SoC,Tem)$ represents the internal impedance/resistance at the beginning of life, $R_{EOL}(SoC,Tem)$ represents the described internal impedance/resistance at the end of life, and $R_{ACT}(SoC,Tem)$ represents the actual internal impedance/resistance under the given (SoC,Tem) conditions.

If the Ahr capacity of a battery is designed as the health index of a battery, the SOL can be defined as follows:

$$SOL_{C\_Based} = \frac{Cap_{ACT}(Tem) - Cap_{EOL}(Tem)}{Cap_{BOL}(Tem) - Cap_{EOL}(Tem)} \times 100\% \quad (3.89)$$

where $Cap_{BOL}(Tem)$ is the Ahr capacity at the beginning of life, $Cap_{EOL}(Tem)$ is the defined Ahr capacity at the end of life, and $Cap_{ACT}(Tem)$ is the actual Ahr capacity at the given temperature.

3.8.2.3.2 SOL Determination under Storage Conditions

To make technical and commercial decisions, it is essential to be able to predict the SOL of a battery based on the storage situations. As mentioned above, the temperature and the SOC affect the life of a battery under storage conditions. In this section, an Arrhenius equation-based SOL calculation is presented. The following improved Arrhenius equation can be applied to describe the age of a battery under storage conditions:

$$G(x) = A(x).\exp\left(-\frac{E_a}{R_g(Tem+273.15)}\right).t_x^{f(SoC)} \tag{3.90}$$

where $G(x)$ represents the aging gauge or aging indicator, normally related to the increment of the battery features, x represents the battery feature, namely, internal resistance, power availability, or Ahr capacity, $A(x)$ represents the pre-exponential constant calculated by the system feature, E_a represents the activation energy (kJ/mol) indicating the energy barrier for the thermal activation procedure, $R_g = 8.314\times10^{-3}$ represents the general gas constant ($kJmol^{-1}K^{-1}$), and t represents the time.

The detailed function of *SoC*, $f(SoC)$, and model parameters must be defined according to the test data before the model in Eq. (3.79) is applied to determine the SOL. Based on the Taylor expansion theorem, the can be $f(SoC)$ described as follows:

$$f(SoC) = a_0 + a_1 SoC + a_2 SoC^2; \qquad 0 < f(SoC) < 1; \qquad 0 < SoC < 1 \tag{3.91}$$

where $A(x)$, E_a, a_0, a_1, and a_2 can be determined by solving the following optimization problem.

Objective function:

$$J = \min\sum_{i=1}^{n}\left(G_i(x) - G_{test\_i}(x)\right)^2$$
$$= \min\sum_{i=1}^{n}\left(\hat{A}(x).\exp\left(-\frac{\hat{E}_a}{R_g(Tem+273.15)}\right).t_x^{\left(\hat{a}_0+\hat{a}_1 SoC+\hat{a}_2 SoC^2\right)} - G_{test\_i}(x)\right)^2 \tag{3.92}$$

Subject to:

$$\begin{aligned}
&A^{low}(x) \le A(x) \le A^{upper}(x) \\
&E^{low} \le E_a \le E^{upper} \\
&a_0^{low} \le a_0 \le a_0^{upper} \\
&a_1^{low} \le a_1 \le a_1^{upper} \\
&a_2^{low} \le a_1 \le a_2^{upper} \\
&0 < a_0 + a_1 SoC + a_2 SoC^2 < 1
\end{aligned} \tag{3.93}$$

where $G_i(x)$ represents the test data i.

Once the model parameters have been assessed, the SOL can be computed from Eq. (3.90). If the aging index is set as the power availability, the SOL of the battery can be assessed based on the storage time as follows:

$$\begin{aligned}
SOL_{P_Based} &= \frac{P_{ACT}(SoC,Tem) - P_{EOL}(SoC,Tem)}{P_{BOL}(SoC,Tem) - P_{EOL}(SoC,Tem)} \times 100\% \\
&= \frac{P_{BOL}(SoC,Tem) - G(P) - P_{EOL}(SoC,Tem)}{P_{BOL}(SoC,Tem) - P_{EOL}(SoC,Tem)} \times 100\% \\
&= \frac{\Delta P^{\max} - G(P)}{\Delta P^{\max}(SoC,Tem)} \times 100\% \\
&= \left(1 - \frac{A(P).\exp\left(-\frac{E_a}{R_g(Tem + 273.15)} \right)}{\Delta P^{\max}(SoC,Tem)} \right) \times 100\%
\end{aligned} \tag{3.94}$$

where $P_{BOL}(SoC,Tem)$ represents the power availability of the battery at the beginning of life, $P_{EOL}(SoC,Tem)$ represents the developed battery power availability at the end of life, and $G(P)$ represents the decline in the power availability under the given storage conditions.

If the aging index is set as the internal resistance, the SOL of a battery can be characterized based on the storage time as follows:

$$SOL_{R\_Based} = \frac{R_{EOL} - R_{ACT}}{R_{EOL} - R_{BOL}} \times 100\% = \frac{R_{EOL} - R_{BOL} - G(P)}{\Delta R^{\max}} \times 100\% = \frac{\Delta R^{\max} - G(P)}{\Delta R^{\max}} \times 100\%$$
$$= \left(1 - \frac{A(R).\exp\left(-\frac{E_a}{R_g(Tem + 273.15)}\right).t_R^{f(SoC)}}{\Delta R^{\max}}\right) \times 100\% \tag{3.95}$$

where R_{BOL} represents the battery's internal resistance at the beginning of life, R_{EOL} represents the designed battery's internal resistance at the end of life, and $G(P)$ represents the increment of the internal resistance during the period, t, under the given storage conditions.

If the aging index is set as the Ahr capacity, the SOL of a battery can be characterized based on the storage time as follows:

$$SOL_{C\_Based} = \frac{Cap_{ACT}(Tem) - Cap_{EOL}(Tem)}{Cap_{BOL}(Tem) - Cap_{EOL}(Tem)} \times 100\% = \frac{Cap_{BOL}(Tem) - G(P) - Cap_{EOL}(Tem)}{Cap_{BOL}(Tem) - Cap_{EOL}(Tem)} \times 100\%$$
$$\frac{\Delta Cap^{\max} - G(Cap)}{\Delta Cap^{\max}} \times 100\% = \left(1 - \frac{A(Cap).\exp\left(-\frac{E_a}{R_g(Tem + 273.15)}\right).t_{Cap}^{f(SoC)}}{\Delta Cap^{\max}}\right) \times 100\% \tag{3.96}$$

where Cap_{BOL} represents the Ahr capacity of the battery at the beginning of life, Cap_{EOL} represents the designed Ahr capacity of the battery at the end of life, and $G(Cap)$ represents the decline in the Ahr capacity under the given storage conditions.

3.8.2.3.3 SOL Determination under Cycling Conditions

Compared with those under storage situations, the aging procedures under cycling conditions are even more complicated. The aging in cycling enhances kinetically induced effects, such as volume variations or concentration gradients.

- Offline Lifetime Determination under Cycling Conditions

The offline lifetime computation requires considering various cycling conditions. According to the calculations, the hybrid vehicle system engineers can size the energy storage system, set system operational points, determine the guarantee period, assess system performance, and compute the general system cost. As mentioned above, the lifetime of a battery under cycling conditions is generally expressed as follows:

$$life\_time = f\left(e^{\alpha.Tem}, SoC_{sp}, \Delta SoC, ATP_{nor}, I_{nor}^2\right) \tag{3.97}$$

where SoC_{sp} represents the operational SoC setpoint, ΔSoC represents the designed SoC swing, ATP_{nor} represents the normalized ampere-hour throughput, and I_{nor}^2 represents the normalized current square based on the battery capacity. ATP_{nor} and I_{nor}^2 are related to the designed annual mileage and cycling intensity. According to the Taylor expansion theorem, the model presented in Eq. (3.97) can be approximated (omitting second-order) as follows:

$$\begin{aligned} life\_time &= f\left(e^{\alpha.Tem}, SoC_{sp}, \Delta SoC, ATP_{nor}, I_{nor}^2\right) \\ life\_time &= a_0 + a_1.Tem + a_2.SoC_{sp} + a_3.\Delta SoC + a_4.ATP_{nor} + a_5.I_{nor}^2 \\ &+ a_6.\left(Tem\right)^2 + a_7.\left(SoC_{sp}\right)^2 + a_8.\left(\Delta SoC\right)^2 + a_9.\left(ATP_{nor}\right)^2 + a_{10}.\left(I_{nor}^2\right)^2 \\ &+ a_{11}.Tem.SoC + a_{12}.Tem.\Delta SoC + a_{13}.Tem.ATP_{nor} + a_{14}.Tem.I_{nor}^2 \\ &+ a_{15}.SoC\Delta SoC + a_{16}.SoC.ATP_{nor} + a_{17}.SoC.I_{nor}^2 \\ &+ a_{18}.\Delta SoC.ATP_{nor} + a_{19}.\Delta SoC.I_{nor}^2 + a_{20}.ATP_{nor}.I_{nor}^2 \end{aligned} \tag{3.98}$$

where the coefficients $a_i \forall i = 0,1,...,20$ are determined by the test data.

Since this model is based on an empiric statistical model, the sample size has to be sufficiently big to provide the accuracy requirements, and the raw data are normally required from practical vehicles owned by users rather than from labs.

- **Online SOL Determination Based on the Estimated Internal Resistance in Real Time**

As defined in the prior section, if the internal resistance of a battery can be determined in real time, the SOL can be estimated in real time by the following equation:

$$SOL_{R\_Based} = \frac{R_{EOL} - R_{ACT}}{R_{EOL} - R_{BOL}} \times 100\% = \frac{R_{EOL} - \breve{R}_{real\_time}}{R_{EOL} - R_{BOL}} \times 100\% \tag{3.99}$$

where $\breve{R}_{real\_time}$ represents the estimated resistance in real-time.

As presented in Section 3.7, the battery terminal voltage can be expressed as:

$$V^T = V^{OC} + V^{Ohm} + V^{Dyn} = V^{OC} + R^{Ohm}.I + V^{Dyn} \tag{3.100}$$

If the battery is modeled by the one *RC* pair electrical circuit shown in Figure 3.6, the terminal voltage can be defined at time *k* as follows:

$$V^T(k) = \alpha_1.V^T(k-1) + \alpha_2.I(k) + \alpha_3.I(k-1) + \alpha_4 \tag{3.101}$$

where

$$\alpha_1 = \left(1 - \frac{\Delta Tem}{R^{Dyn}C^{Dyn}}\right); \quad \alpha_2 = R^{Ohm}; \quad \alpha_3 = \frac{(R^{Dyn} + R^{Ohm})\Delta Tem}{R^{Dyn}C^{Dyn}} - R^{Ohm}; \quad \alpha_4 = \frac{\Delta Tem.V^{OC}}{R^{Dyn}C^{Dyn}}$$

The following relations sustain between the parameters of the difference Eq. (3.101) and the physical parameters of the battery model:

$$R^{Ohm} = \alpha_2; \quad V^{OC} = \frac{\alpha_4}{1-\alpha_1}; \quad R^{Dyn} = \frac{\alpha_3 + \alpha_1.\alpha_2}{1-\alpha_1}; \quad C^{Dyn} = \frac{\Delta Tem}{\alpha_3 + \alpha_1.\alpha_2} \tag{3.102}$$

The parameters α_1, α_2, α_3, and α_4 in Eq. (3.101) can be calculated in real-time according to the measurement of battery terminal voltage via a recursive estimation approach and obtained the ohmic resistance in real time. In practice, a moving average filter is normally applied to further smooth the estimated resistance $R^{Ohm}k$ over the drive cycle as follows:

$$\overline{R}^{Ohm} = \frac{\sum_{i=1}^{N} \hat{R}^{Ohm}(i)}{N} \tag{3.103}$$

After obtaining R^{Ohm}, the SOL can be calculated based on Eq. (3.99) in real time [7, 79].

3.8.3 RUL Prediction Methods

Based on the principles and feasible conditions of the methods applied, this section provides the methods applied in many research works and divides the RUL prediction methods into artificial intelligence-based, filtering-based, and statistical data-driven methods. These methods can introduce connections and trends corresponding to degradation according to the available data. The RUL prediction methods can be divided into three main strategies as follows below.

3.8.3.1 Machine Learning Methods

The artificial neural network has been extensively applied for self-learning and self-organization, and it does not depend on the electrochemical principles of the battery. Machine learning algorithms can learn and realize more complex data patterns in various utilizations according to experience. Machine learning is applied for forecasting by gathering historical data in the life cycle. One of the principal methods of machine learning is neural network strategies. The neural network can include several layers, and each layer can also include several neurons. Synthesis of incremental capacity analysis (ICA) and a radial-based function neural network (RBFNN) model are applied to evaluate battery aging. Establishment based on the electric city bus operation data set demonstrated that the average prediction error of this strategy reaches 4%, the confidence interval of the derived model is 92%, and the prediction accuracy is 90% [80].

3.8.3.2 Adaptive Filter Methods

The adaptive filter is a digital filter, and its coefficient alternations with the target to make the filter converge to the best state. The optimization objective is the regular cost function, and the most general is the root mean square of the error signal between the adaptive filter output and the required signal. When you change the input data characteristics, the filter modifies the current environment by generating an up-to-date set of coefficients for the latest data. The adaptive filter leads to a prompt prediction of the system state according to the confidence interval [81].

3.8.3.3 Stochastic Process Methods

Stochastic process methods rely on statistical theory and are integrated with other mathematical principles. Statistics-driven methods are commonly divided into three categories including Bayesian estimation, ground-penetrating radar (GPR), and wiener process (WP). Bayesian estimation requires the utilization of posterior prediction distributions for predictive inference, predicting the new distribution, and unobserved data points. Contrary to a fixed point as a prediction, return the possible distribution points [82]. GPR is suitable for handling complex regression issues, such as enhanced dimensions, few samples, and nonlinearity [83]. WP is a typical random procedure, which refers to the so-called independent incremental procedure. It begins with the theory and utilization of random procedures. WP can explain not only the monotonic degradation of equipment efficiency but also the non-monotonic degradation of equipment [84].

3.9 Recent Trends, Future Extensions, and Challenges of ESSs in EV Implementations

Electric mobility is a progressive trend in the automotive market, which provides a requirement for the ESSs all over the world. Government agencies such as environmental agencies presently focus on the descriptions of boundaries on carbon dioxide emission. For example, China has a primary objective to enhance the production of electric cars. China's focus is to establish almost 18.7 million electric cars units by 2022. There is further a universal focus to reduce carbon dioxide emission and provide EVs, which will be environment-friendly and ensure transportation requirements [85]. The recent improvement of ESSs is desirable for the ESSs applied in EV. The ESSs are regularly mature with technological alternations and developments in certain applications. However, these applications still endure issues including raw material sustain and suitable disposal, energy control, power electronics interface, sizing, safety factors, and expenses. The mentioned problems of current ESSs are the main challenges to advanced investigations for the development of ESSs in EV implementations. The main challenges are described in the following sections.

- **Materials**
  The accessibility of raw materials for producing the ESSs and the advancement of the related products is a key challenge. The electrodes, electrolyte, partitioner materials, and chemical solutions, flywheel materials, superconducting materials for SMES, and hydrogen fuel for FCs are vital in ESS manufacturing. The main development in the ESS technology for EV implementations is to take into account the optimization of the ESS materials, alloys, and solution preparation, and also the utilization of the ESSs with high power and energy density, proper charge/discharge rate profiles, lifetime, cost, and secure operation without corrosion [86, 87].

- **Power Electronics**
  All energy resources and ESSs can generate and supply power based on their natural properties regardless of performance and optimal supply of voltage and current. Undesignated and unsystematic power storage could decrease the efficiency and life cycle period of the ESS, and also cause high power loss, unusual damages, and constrained manner of connected loads. The power electronics interface addresses situations to deal with power conditions, management, and conversions for saving and returning the saved power of ESS and load demands such that the total efficiency and lifetime of that system are optimized. The ESSs in EV implementations require a power electronics interface for power flow control, power control, motor drive, charging balance, and secure operation. Furthermore, the hybridization of ESSs to provide proper, efficient, and balanced power and energy generation needs power electronics converters for EV motor systems. Notably, the current power electronics interfaces endure either in implementation scale, power efficiency, voltage stress, reliability, or expenses. Therefore, innovative investigations are required in the field of power electronics to control power supply and efficiency by decreasing the corresponding losses [88–90].

- **Energy Control**
  The EVs normally operate on generated power from batteries, UCs, FCs, and hybrid ESSs. An energy management system (EMS) controls all available energy resources to supply the power to ESSs in EVs. The EMS handles energy resource systems, energy storage, and power electronics. The accessible energy resources for recharging of the ESSs in EV are mainly power networks, regenerative braking, solar energy, hydrogen energy, thermal energy, flywheels, and other similar energy sources. The advanced EV systems are developed to provide the control of all energy resources efficiently in such a way that power availability and loads using EMS could optimize energy economy and performance [91, 92].

- **Implementation Scale and Expenses**
  Compact mechanical energy storages normally designed with an enhanced capacity and economic composition are the major challenges for the improvement of next-generation EVs. About 30% of the overall expenses of EV formation are assigned to the implemented ESS even though the expenses differ for diverse ESS technologies. The overall expenses of ESSs contain the cost of raw materials, power adaptation, operation and maintenance, and employment. It can be mentioned that electrochemical SBs, namely, CAES, UCs, and PHS have reduced overall expense per unit energy, while flywheels and SMESs have an enhanced energy storage expense. Besides, batteries and FCs have an enhanced power expense per unit, even though such property is reduced in SMES and UCs. Therefore, the expenses of EESs for EV implementations could be calculated by getting the ideas from this extensive storage cost overview [93–95].

- **Ecological Aspects**
  Despite the application of the EV concept and the decreased fuel demand, the ESSs applied in EV have had small influence on environmental pollution in the procedure of producing, discarding, and recycling of ESSs, particularly electrochemical batteries. Furthermore, the manufacturing of ESS leads to unpleasant pulmonary, respiratory, and neurological causes. Hence, safety factors and advanced appliances are vital to deal with the entire procedure of

manufacturing and maintaining ESSs, particularly in EV implementations [96, 97].

- **Safety Factors**
  Safety factors guarantee the effective function of ESSs and increment of their efficiencies. Every ESS needs protection and periodical maintenance for its effective function. For EV implementations, the Li-ion batteries should be protected from over-charging and over-discharging, the Zn-Air batteries require short circuit protection, the Na-S batteries demand high-temperature safety control, the ZEBRA batteries require a thermal control, the FCs require low- and high-temperature maintenance schemes, and the LA batteries need periodical. Advanced EVs apply power electronics interfaces as energy and power control and power conversion to provide effectual services and security functions of the ESSs [98, 99].

- **Market Trends**
  The key players in the global ESSs market adopt diverse key business methods such as expansion, agreement, product launch, acquisition, partnership, collaboration, and product improvements to sustain in the global competitive market. The market players can be contributed as manufacturers and solution providers of ESSs. The transportation section is expected to increase exponentially during the predicted period, due to the increasing adoption of fuel cells in EVs, as they enhance vehicle performance. The batteries such as Li-ion and lead-acid also sustain market potential for enhancement considering the focus of main economies such as China to develop EVs [100]. Key trends that lead to the market integration of ESSs can be described as follows:

  - Enhanced adoption of ESSs in transportation application
  - Software combination for ESS management
  - Enhanced adoption of renewable energies

3.10 Government Policy Challenges for EVs

The convolution of altering the human transportation conception involves further technical and economic factors. Recent technologies like EVs meet various notable barriers to wider adoption, including the high purchase cost, the lack of charging infrastructure, limited driving range, and long charging period. However, EV owners can charge their vehicles overnight at home. Considering the limited driving range, users still have to worry about running out of charge before reaching their destination. This matter of range anxiety could lead to reluctance to adopt EVs, particularly when public charging stations are difficult to access. Simultaneously, private investors have less motivation to set up charging stations if the size of the EV fleet and market potential is small. The collaboration between both sides of the market (charging stations and EVs) can be taken into account as indirect network results; the advantage of investment/adoption on one side of the market develops with the network size of the other side of the market. The International Energy Agency predicts that by 2050, EVs have the prospect to account for half of the light-duty vehicle sales. Several countries throughout the world have expanded purposes to enhance the EV market and provide the required supports to expand the diffusion of this technology. To reduce the price gap between electric and gasoline vehicles, the Energy Reform and Extension Act of 2008, and then the American Clean Energy and Security Act of 2009, give a federal tax credit for novel qualified EVs. The minimum credit is determined $2,000 up to $8,000, based on the battery capacity of the EVs. Furthermore, many states have appointed extra state-level incentives to better enhance EV adoption like tax exemptions and rebates for EVs, and nonmonetary incentives including toll reduction, high occupancy vehicle lane access, and free parking [101].

Government intervention in this market could be well founded from the few prospects. First, indirect network impacts in the EV market imply a market failure since marginal users/stockholders only take into account private profit in their decision, and the network size on both sides is less than optimal. Furthermore, according to the nature of the market, each market side is unlikely to internalize the external aspects on the other side through market transactions. If EVs are manufactured by an automaker, it would have a motivation to provide a charging station network to enhance EV adoption. Second, the external costs from gasoline utilization in many countries throughout the world are not properly reflected by the gasoline taxes. In comparison with conventional gasoline vehicles, EVs offer environmental profits when the electricity comes from renewable energies such as photovoltaics and wind turbines. Electricity generation continues to become cleaner energy throughout the world because of the adoption of abatement technologies, the propagation of renewable energies, and the alternation from coal to natural gas. The integration of the RES-based units with EV charging not only can assist EVs in fully realize their environmental advantages but also can help EV batteries to address the issues corresponding to the intermittency [102].

3.11 Conclusion

EV technologies are substitute decisions for gasoline-based transportations considering that regular vehicles with gasoline fuel provide reduced drive train performance and CO2 emissions. The request for EVs is significantly enhancing according to the zero-emission principles. Therefore, the refinement in the development of EVs with intelligent establishments is a challenging issue for next-generation EVs. However, efficient EVs cannot be designed without taking into account the novel technologies of ESSs. This chapter aimed to review and model the existing ESSs, explain the corresponding equations, and energy transformation procedure with the diverse characteristics for ESSs in EVs. A detailed analysis of all EV aspects is provided with a focus on the recent development and design of EV batteries. The approaches of calculating the SOC, the estimation of battery power availability, and the life prediction of the battery were given comprehensively. Besides, the characteristics and application of ESSs in EVs were discussed by investigating the ESS technologies, life cycles, and performances. This chapter also addressed recent trends, future extensions, and challenges of ESS technologies in EV implementations. Typically, the majority of the feasible technologies of the ESSs included in EVs are following the electrochemical batteries or FCs. Besides, the hybridization of ESSs with reciprocal features has been investigated to ESSs technologies in EV for developing ESS properties. These properties contain energy and power density, response time, performance, life cycle, and expenses balancing. Therefore, the ESSs, in particular, flywheel, SMES, and UC are presented as systems for providing a hybrid ESS. Because of the recent developments in Li-ion and NiMH battery technologies, the electrified transportation system can meet their main objectives. The novel methodologies including high-capacity anode material, high-capacity metal oxide cathode materials, and new electrolytes with high oxidation capacity, the metal-air batteries after alternating the positive electrode with an air electrode may assist the batteries of EVs to effectively perform.

The ESSs are progressively becoming fully developed with technological alternations and improvements for particular investigations. However, these ESSs still have some issues including energy management, proper recycling, material support, power electronics interface, safety measures, sizing, and investments. The optimization of high-grade ESS materials, alloys, and chemical solutions in the design of ESSs for EV applications could be taken into account for developing charge capacity, power, and energy density, recharge/discharge rate profiles, and environmental and economic effects. Modern EV implementations could control available energy resources effectively to save the possible amount of energy and supply power to network demands via an appropriate EMS to optimize power efficiency and energy expenses. Finally, it can be concluded that the next-generation EVs would control ESS to save required energy and to drive itself, besides, to become an emergency mobile power backup and provide V2G facility concerning fast improvements for EVs.

References

1. S. E. Ahmadi and N. Rezaei, "A new isolated renewable based multi microgrid optimal energy management system considering uncertainty and demand response," *International Journal of Electrical Power & Energy Systems*, vol. 118, p. 105760, 2020/06/01/ 2020, doi: https://doi.org/10.1016/j.ijepes.2019.105760.
2. S. E. Ahmadi, N. Rezaei, and H. Khayyam, "Energy management system of networked microgrids through optimal reliability-oriented day-ahead self-healing scheduling," *Sustainable Energy, Grids and Networks*, vol. 23, p. 100387, 2020/09/01/ 2020, doi: https://doi.org/10.1016/j.segan.2020.100387.
3. Y. Balali and S. Stegen, "Review of energy storage systems for vehicles based on technology, environmental impacts, and costs," *Renewable and Sustainable Energy Reviews*, vol. 135, p. 110185, 2021/01/01/ 2021, doi: https://doi.org/10.1016/j.rser.2020.110185.
4. C. Zheng, W. Li, and Q. Liang, "An Energy Management Strategy of Hybrid Energy Storage Systems for Electric Vehicle Applications," *IEEE Transactions on Sustainable Energy*, vol. 9, no. 4, pp. 1880-1888, 2018, doi: 10.1109/TSTE.2018.2818259.
5. Y. Li, X. Huang, D. Liu, M. Wang, and J. Xu, "Hybrid energy storage system and energy distribution strategy for four-wheel independent-drive electric vehicles," *Journal of Cleaner Production*, vol. 220, pp. 756-770, 2019/05/20/ 2019, doi: https://doi.org/10.1016/j.jclepro.2019.01.257.

6. S. Koohi-Fayegh and M. A. Rosen, "A review of energy storage types, applications and recent developments," *Journal of Energy Storage,* vol. 27, p. 101047, 2020/02/01/ 2020, doi: https://doi.org/10.1016/j.est.2019.101047.
7. W. Liu, *Hybrid electric vehicle system modeling and control.* John Wiley & Sons, 2017.
8. M. S. Guney and Y. Tepe, "Classification and assessment of energy storage systems," *Renewable and Sustainable Energy Reviews,* vol. 75, pp. 1187-1197, 2017/08/01/ 2017, doi: https://doi.org/10.1016/j.rser.2016.11.102.
9. M. Aneke and M. Wang, "Energy storage technologies and real life applications – A state of the art review," *Applied Energy,* vol. 179, pp. 350-377, 2016/10/01/ 2016, doi: https://doi.org/10.1016/j.apenergy.2016.06.097.
10. H. S. Das, M. M. Rahman, S. Li, and C. W. Tan, "Electric vehicles standards, charging infrastructure, and impact on grid integration: A technological review," *Renewable and Sustainable Energy Reviews,* vol. 120, p. 109618, 2020/03/01/ 2020, doi: https://doi.org/10.1016/j.rser.2019.109618.
11. S. Sharma, A. K. Panwar, and M. M. Tripathi, "Storage technologies for electric vehicles," *Journal of Traffic and Transportation Engineering (English Edition),* vol. 7, no. 3, pp. 340-361, 2020/06/01/ 2020, doi: https://doi.org/10.1016/j.jtte.2020.04.004.
12. L. Kouchachvili, W. Yaïci, and E. Entchev, "Hybrid battery/supercapacitor energy storage system for the electric vehicles," *Journal of Power Sources,* vol. 374, pp. 237-248, 2018/01/15/ 2018, doi: https://doi.org/10.1016/j.jpowsour.2017.11.040.
13. R. Machlev, N. Zargari, N. R. Chowdhury, J. Belikov, and Y. Levron, "A review of optimal control methods for energy storage systems - energy trading, energy balancing and electric vehicles," *Journal of Energy Storage,* vol. 32, p. 101787, 2020/12/01/ 2020, doi: https://doi.org/10.1016/j.est.2020.101787.
14. B. Xiao, J. Ruan, W. Yang, P. D. Walker, and N. Zhang, "A review of pivotal energy management strategies for extended range electric vehicles," *Renewable and Sustainable Energy Reviews,* vol. 149, p. 111194, 2021/10/01/ 2021, doi: https://doi.org/10.1016/j.rser.2021.111194.
15. L. Zhang *et al.*, "Hybrid electrochemical energy storage systems: An overview for smart grid and electrified vehicle applications," *Renewable and Sustainable Energy Reviews,* vol. 139, p. 110581, 2021/04/01/ 2021, doi: https://doi.org/10.1016/j.rser.2020.110581.
16. A. Ibrahim and F. Jiang, "The electric vehicle energy management: An overview of the energy system and related modeling and simulation," *Renewable and Sustainable Energy Reviews,* vol. 144, p. 111049, 2021/07/01/ 2021, doi: https://doi.org/10.1016/j.rser.2021.111049.
17. M. A. Hannan, M. M. Hoque, A. Mohamed, and A. Ayob, "Review of energy storage systems for electric vehicle applications: Issues and challenges," *Renewable and Sustainable Energy Reviews,* vol. 69, pp. 771-789, 2017/03/01/ 2017, doi: https://doi.org/10.1016/j.rser.2016.11.171.

18. N. Shaukat *et al.*, "A survey on electric vehicle transportation within smart grid system," *Renewable and Sustainable Energy Reviews,* vol. 81, pp. 1329-1349, 2018/01/01/ 2018, doi: https://doi.org/10.1016/j.rser.2017.05.092.
19. A. Damiano, G. Gatto, I. Marongiu, M. Porru, and A. Serpi, "The Plug-in Electric Vehicles Role in Smart Grid Development: a Survey," in *European Electric Vehicle Congress (EEVC 2012)*, 2012.
20. T. U. Solanke, V. K. Ramachandaramurthy, J. Y. Yong, J. Pasupuleti, P. Kasinathan, and A. Rajagopalan, "A review of strategic charging–discharging control of grid-connected electric vehicles," *Journal of Energy Storage,* vol. 28, p. 101193, 2020/04/01/ 2020, doi: https://doi.org/10.1016/j.est.2020.101193.
21. M. Mubarak, H. Üster, K. Abdelghany, and M. Khodayar, "Strategic network design and analysis for in-motion wireless charging of electric vehicles," *Transportation Research Part E: Logistics and Transportation Review,* vol. 145, p. 102179, 2021/01/01/ 2021, doi: https://doi.org/10.1016/j.tre.2020.102179.
22. P. Machura and Q. Li, "A critical review on wireless charging for electric vehicles," *Renewable and Sustainable Energy Reviews,* vol. 104, pp. 209-234, 2019/04/01/ 2019, doi: https://doi.org/10.1016/j.rser.2019.01.027.
23. S. Niu, H. Xu, Z. Sun, Z. Y. Shao, and L. Jian, "The state-of-the-arts of wireless electric vehicle charging via magnetic resonance: principles, standards and core technologies," *Renewable and Sustainable Energy Reviews,* vol. 114, p. 109302, 2019/10/01/ 2019, doi: https://doi.org/10.1016/j.rser.2019.109302.
24. M. Gholami, H. Tarimoradi, N. Rezaei, A. Ahmadi, and S. E. Ahmadi, "Determining the Type and Size of Energy Storage Systems to Smooth the Power of Renewable Energy Resources," in *Integration of Clean and Sustainable Energy Resources and Storage in Multi-Generation Systems: Design, Modeling and Robust Optimization*, F. Jabari, B. Mohammadi-Ivatloo, and M. Mohammadpourfard Eds. Cham: Springer International Publishing, 2020, pp. 29-59.
25. M. Mahmoud, M. Ramadan, A.-G. Olabi, K. Pullen, and S. Naher, "A review of mechanical energy storage systems combined with wind and solar applications," *Energy Conversion and Management,* vol. 210, p. 112670, 2020/04/15/ 2020, doi: https://doi.org/10.1016/j.enconman.2020.112670.
26. S. M. Mousavi G, F. Faraji, A. Majazi, and K. Al-Haddad, "A comprehensive review of Flywheel Energy Storage System technology," *Renewable and Sustainable Energy Reviews,* vol. 67, pp. 477-490, 2017/01/01/ 2017, doi: https://doi.org/10.1016/j.rser.2016.09.060.
27. K. S. Gallagher and E. Muehlegger, "Giving green to get green? Incentives and consumer adoption of hybrid vehicle technology," *Journal of Environmental Economics and Management,* vol. 61, no. 1, pp. 1-15, 2011/01/01/ 2011, doi: https://doi.org/10.1016/j.jeem.2010.05.004.
28. H. Wu and D. Niu, "Study on Influence Factors of Electric Vehicles Charging Station Location Based on ISM and FMICMAC," *Sustainability,* vol. 9, no. 4, p. 484, 2017. [Online]. Available: https://www.mdpi.com/2071-1050/9/4/484.

29. J. F. Mike and J. L. Lutkenhaus, "Electrochemically Active Polymers for Electrochemical Energy Storage: Opportunities and Challenges," *ACS Macro Letters,* vol. 2, no. 9, pp. 839-844, 2013/09/17 2013, doi: 10.1021/mz400329j.
30. X. Wang, J. Chai, and J. J. Jiang, "Redox flow batteries based on insoluble redox-active materials. A review," *Nano Materials Science,* vol. 3, no. 1, pp. 17-24, 2021/03/01/ 2021, doi: https://doi.org/10.1016/j.nanoms.2020.06.003.
31. J. Sun, M. Wu, H. Jiang, X. Fan, and T. Zhao, "Advances in the design and fabrication of high-performance flow battery electrodes for renewable energy storage," *Advances in Applied Energy,* vol. 2, p. 100016, 2021/05/26/ 2021, doi: https://doi.org/10.1016/j.adapen.2021.100016.
32. K. T. Chau, Y. S. Wong, and C. C. Chan, "An overview of energy sources for electric vehicles," *Energy Conversion and Management,* vol. 40, no. 10, pp. 1021-1039, 1999/07/01/ 1999, doi: https://doi.org/10.1016/S0196-8904(99)00021-7.
33. A. R. Mainar, E. Iruin, and J. A. Blázquez, "High performance secondary zinc-air/silver hybrid battery," *Journal of Energy Storage,* vol. 33, p. 102103, 2021/01/01/ 2021, doi: https://doi.org/10.1016/j.est.2020.102103.
34. P. Křivík, P. Bača, and J. Kazelle, "Effect of ageing on the impedance of the lead-acid battery," *Journal of Energy Storage,* vol. 36, p. 102382, 2021/04/01/ 2021, doi: https://doi.org/10.1016/j.est.2021.102382.
35. A. Calborean, T. Murariu, and C. Morari, "Optimized lead-acid grid architectures for automotive lead-acid batteries: An electrochemical analysis," *Electrochimica Acta,* vol. 372, p. 137880, 2021/03/10/ 2021, doi: https://doi.org/10.1016/j.electacta.2021.137880.
36. G. Li *et al.*, "Advanced intermediate temperature sodium–nickel chloride batteries with ultra-high energy density," *Nature Communications,* vol. 7, no. 1, p. 10683, 2016/02/11 2016, doi: 10.1038/ncomms10683.
37. X. Zhao and L. Ma, "Recent progress in hydrogen storage alloys for nickel/metal hydride secondary batteries," *International Journal of Hydrogen Energy,* vol. 34, no. 11, pp. 4788-4796, 2009/06/01/ 2009, doi: https://doi.org/10.1016/j.ijhydene.2009.03.023.
38. S. K. Dhar, S. R. Ovshínsky, P. R. Gifford, D. A. Corrigan, M. A. Fetcenko, and S. Venkatesan, "Nickel/metal hydride technology for consumer and electric vehicle batteries—a review and up-date," *Journal of Power Sources,* vol. 65, no. 1, pp. 1-7, 1997/03/01/ 1997, doi: https://doi.org/10.1016/S0378-7753(96)02599-2.
39. A. Khor *et al.*, "Review of zinc-based hybrid flow batteries: From fundamentals to applications," *Materials Today Energy,* vol. 8, pp. 80-108, 2018/06/01/ 2018, doi: https://doi.org/10.1016/j.mtener.2017.12.012.
40. K. S. Archana *et al.*, "Effect of positive electrode modification on the performance of zinc-bromine redox flow batteries," *Journal of Energy Storage,* vol. 29, p. 101462, 2020/06/01/ 2020, doi: https://doi.org/10.1016/j.est.2020.101462.

41. H.-F. Wang and Q. Xu, "Materials Design for Rechargeable Metal-Air Batteries," *Matter,* vol. 1, no. 3, pp. 565-595, 2019/09/04/ 2019, doi: https://doi.org/10.1016/j.matt.2019.05.008.
42. X. Chen *et al.*, "A review on recent advancement of nano-structured-fiber-based metal-air batteries and future perspective," *Renewable and Sustainable Energy Reviews,* vol. 134, p. 110085, 2020/12/01/ 2020, doi: https://doi.org/10.1016/j.rser.2020.110085.
43. C. Lin *et al.*, "High-voltage asymmetric metal–air batteries based on polymeric single-Zn2+-ion conductor," *Matter,* vol. 4, no. 4, pp. 1287-1304, 2021/04/07/ 2021, doi: https://doi.org/10.1016/j.matt.2021.01.004.
44. X. Lu, G. Xia, J. P. Lemmon, and Z. Yang, "Advanced materials for sodium-beta alumina batteries: Status, challenges and perspectives," *Journal of Power Sources,* vol. 195, no. 9, pp. 2431-2442, 2010/05/01/ 2010, doi: https://doi.org/10.1016/j.jpowsour.2009.11.120.
45. P. Kehne, C. Guhl, L. Alff, R. Hausbrand, and P. Komissinskiy, "The effect of calcium impurities of β-alumina on the degradation of NaxCoO2 cathodes in all solid state sodium-ion batteries," *Solid State Ionics,* vol. 341, p. 115041, 2019/11/05/ 2019, doi: https://doi.org/10.1016/j.ssi.2019.115041.
46. S. Chen, F. Feng, Y. Yin, H. Che, X.-Z. Liao, and Z.-F. Ma, "A solid polymer electrolyte based on star-like hyperbranched β-cyclodextrin for all-solid-state sodium batteries," *Journal of Power Sources,* vol. 399, pp. 363-371, 2018/09/30/ 2018, doi: https://doi.org/10.1016/j.jpowsour.2018.07.096.
47. S. Liu, H. Zhang, and X. Xu, "A study on the transient heat generation rate of lithium-ion battery based on full matrix orthogonal experimental design with mixed levels," *Journal of Energy Storage,* vol. 36, p. 102446, 2021/04/01/ 2021, doi: https://doi.org/10.1016/j.est.2021.102446.
48. L. Xia, H. Miao, C. Zhang, G. Z. Chen, and J. Yuan, "Review—recent advances in non-aqueous liquid electrolytes containing fluorinated compounds for high energy density lithium-ion batteries," *Energy Storage Materials,* vol. 38, pp. 542-570, 2021/06/01/ 2021, doi: https://doi.org/10.1016/j.ensm.2021.03.032.
49. J. Li, F. Li, L. Zhang, H. Zhang, U. Lassi, and X. Ji, "Recent Applications of Ionic Liquids in Quasi-Solid-State Lithium Metal Batteries," *Green Chemical Engineering,* 2021/03/10/ 2021, doi: https://doi.org/10.1016/j.gce.2021.03.001.
50. L. Zhang, H. B. Wu, and X. W. Lou, "Iron-Oxide-Based Advanced Anode Materials for Lithium-Ion Batteries," *Advanced Energy Materials,* https://doi.org/10.1002/aenm.201300958 vol. 4, no. 4, p. 1300958, 2014/03/01 2014, doi: https://doi.org/10.1002/aenm.201300958.
51. W. Huang *et al.*, "Colloidal to micrometer-sized iron oxides and oxyhydroxides as anode materials for batteries and pseudocapacitors: Electrochemical properties," *Colloids and Surfaces A: Physicochemical and Engineering Aspects,* vol. 615, p. 126232, 2021/04/20/ 2021, doi: https://doi.org/10.1016/j.colsurfa.2021.126232.

52. G. Alva, Y. Lin, and G. Fang, "An overview of thermal energy storage systems," *Energy,* vol. 144, pp. 341-378, 2018/02/01/ 2018, doi: https://doi.org/10.1016/j.energy.2017.12.037.
53. S. Li *et al.*, "Designing interfacial chemical bonds towards advanced metal-based energy-storage/conversion materials," *Energy Storage Materials,* vol. 32, pp. 477-496, 2020/11/01/ 2020, doi: https://doi.org/10.1016/j.ensm.2020.07.023.
54. C. L. Aardahl and S. D. Rassat, "Overview of systems considerations for on-board chemical hydrogen storage," *International Journal of Hydrogen Energy,* vol. 34, no. 16, pp. 6676-6683, 2009/08/01/ 2009, doi: https://doi.org/10.1016/j.ijhydene.2009.06.009.
55. J. E. Valdez-Resendiz, J. C. Rosas-Caro, J. C. Mayo-Maldonado, A. Claudio-Sanchez, O. Ruiz-Martinez, and V. M. Sanchez, "Improvement of ultracapacitors-energy usage in fuel cell based hybrid electric vehicle," *International Journal of Hydrogen Energy,* vol. 45, no. 26, pp. 13746-13756, 2020/05/11/ 2020, doi: https://doi.org/10.1016/j.ijhydene.2019.12.201.
56. N. P, S. V. K. N, and S. S. K. V, "Mathematical modeling and stability analysis of an ultracapacitor based energy storage system considering non-idealities," *Journal of Energy Storage,* vol. 33, p. 102112, 2021/01/01/ 2021, doi: https://doi.org/10.1016/j.est.2020.102112.
57. A. Kuperman and I. Aharon, "Battery–ultracapacitor hybrids for pulsed current loads: A review," *Renewable and Sustainable Energy Reviews,* vol. 15, no. 2, pp. 981-992, 2011/02/01/ 2011, doi: https://doi.org/10.1016/j.rser.2010.11.010.
58. P. Mukherjee and V. V. Rao, "Design and development of high temperature superconducting magnetic energy storage for power applications - A review," *Physica C: Superconductivity and its Applications,* vol. 563, pp. 67-73, 2019/08/15/ 2019, doi: https://doi.org/10.1016/j.physc.2019.05.001.
59. A. Z. Al Shaqsi, K. Sopian, and A. Al-Hinai, "Review of energy storage services, applications, limitations, and benefits," *Energy Reports,* vol. 6, pp. 288-306, 2020/12/01/ 2020, doi: https://doi.org/10.1016/j.egyr.2020.07.028.
60. A. Gautam and R. P. Saini, "A review on sensible heat based packed bed solar thermal energy storage system for low temperature applications," *Solar Energy,* vol. 207, pp. 937-956, 2020/09/01/ 2020, doi: https://doi.org/10.1016/j.solener.2020.07.027.
61. A. Mehari, Z. Y. Xu, and R. Z. Wang, "Thermal energy storage using absorption cycle and system: A comprehensive review," *Energy Conversion and Management,* vol. 206, p. 112482, 2020/02/15/ 2020, doi: https://doi.org/10.1016/j.enconman.2020.112482.
62. Y. Jiang, B. Xia, X. Zhao, T. Nguyen, C. Mi, and R. A. de Callafon, "Identification of Fractional Differential Models for Lithium-ion Polymer Battery Dynamics," *IFAC-PapersOnLine,* vol. 50, no. 1, pp. 405-410, 2017/07/01/ 2017, doi: https://doi.org/10.1016/j.ifacol.2017.08.184.

63. S. Parakkadavath and B. Bhikkaji, "Identification of the non-linear dynamics and state of charge estimation of a LiFePO4 battery using constrained unscented Kalman filter," *IFAC-PapersOnLine,* vol. 50, no. 1, pp. 1571-1576, 2017/07/01/ 2017, doi: https://doi.org/10.1016/j.ifacol.2017.08.311.
64. N. Kularatna and K. Gunawardane, "4 - Dynamics, models, and management of rechargeable batteries," in *Energy Storage Devices for Renewable Energy-Based Systems (Second Edition)*, N. Kularatna and K. Gunawardane Eds. Boston: Academic Press, 2021, pp. 99-172.
65. S. Xiang, G. Hu, R. Huang, F. Guo, and P. Zhou, "Lithium-Ion Battery Online Rapid State-of-Power Estimation under Multiple Constraints," *Energies,* vol. 11, no. 2, 2018, doi: 10.3390/en11020283.
66. M. Akbarzadeh *et al.*, "Thermal modeling of a high-energy prismatic lithium-ion battery cell and module based on a new thermal characterization methodology," *Journal of Energy Storage,* vol. 32, p. 101707, 2020/12/01/ 2020, doi: https://doi.org/10.1016/j.est.2020.101707.
67. S. Ayche, M. Daboussy, and E. Aglzim, "Modeling and experimenting the thermal behavior of a lithium-ion battery on a electric vehicle," in *2018 Third International Conference on Electrical and Biomedical Engineering, Clean Energy and Green Computing (EBECEGC)*, 25-27 April 2018 2018, pp. 16-22, doi: 10.1109/EBECEGC.2018.8357126.
68. M. Zhang and X. Fan, "Review on the State of Charge Estimation Methods for Electric Vehicle Battery," *World Electric Vehicle Journal,* vol. 11, no. 1, 2020, doi: 10.3390/wevj11010023.
69. X. L. Liu, Z. M. Cheng, F. Y. Yi, and T. Y. Qiu, "SOC calculation method based on extended Kalman filter of power battery for electric vehicle," in *2017 12th International Conference on Intelligent Systems and Knowledge Engineering (ISKE)*, 24-26 Nov. 2017 2017, pp. 1-4, doi: 10.1109/ISKE.2017.8258840.
70. S. Piller, M. Perrin, and A. Jossen, "Methods for state-of-charge determination and their applications," *Journal of Power Sources,* vol. 96, no. 1, pp. 113-120, 2001/06/01/ 2001, doi: https://doi.org/10.1016/S0378-7753(01)00560-2.
71. W.-Y. Chang, "The state of charge estimating methods for battery: A review," *International Scholarly Research Notices,* vol. 2013, 2013.
72. F. Quantmeyer and X. B. Liu-Henke, "Modeling the Electrical Behavior of Lithium-Ion Batteries for Electric Vehicles," *Solid State Phenomena,* vol. 214, pp. 40-47, 2014, doi: 10.4028/www.scientific.net/SSP.214.40.
73. B. Rzepka, S. Bischof, and T. Blank, "Implementing an Extended Kalman Filter for SoC Estimation of a Li-Ion Battery with Hysteresis: A Step-by-Step Guide," *Energies,* vol. 14, no. 13, 2021, doi: 10.3390/en14133733.
74. K. Saqli, H. Bouchareb, M. Oudghiri, and N. M'sirdi, "An overview of State of Charge (SOC) and State of Health (SOH) estimation methods of Li-ion batteries," in *IMAACA 2019*, 2019.
75. T. Kim, Y. Wang, Z. Sahinoglu, T. Wada, S. Hara, and W. Qiao, "State of charge estimation based on a realtime battery model and iterative smooth variable

structure filter," in *2014 IEEE Innovative Smart Grid Technologies-Asia (ISGT ASIA)*, 2014: IEEE, pp. 132-137.

76. A. Farmann and D. U. Sauer, "A comprehensive review of on-board State-of-Available-Power prediction techniques for lithium-ion batteries in electric vehicles," *Journal of Power Sources*, vol. 329, pp. 123-137, 2016/10/15/ 2016, doi: https://doi.org/10.1016/j.jpowsour.2016.08.031.
77. P. Zhang, J. Liang, and F. Zhang, "An Overview of Different Approaches for Battery Lifetime Prediction," *IOP Conference Series: Materials Science and Engineering*, vol. 199, p. 012134, 2017/05 2017, doi: 10.1088/1757-899x/199/1/012134.
78. S. Wang, S. Jin, D. Deng, and C. Fernandez, "A Critical Review of Online Battery Remaining Useful Lifetime Prediction Methods," (in English), *Frontiers in Mechanical Engineering*, Review vol. 7, no. 71, 2021-August-03 2021, doi: 10.3389/fmech.2021.719718.
79. D. Gao, Y. Zhou, T. Wang, and Y. Wang, "A Method for Predicting the Remaining Useful Life of Lithium-Ion Batteries Based on Particle Filter Using Kendall Rank Correlation Coefficient," *Energies*, vol. 13, no. 16, 2020, doi: 10.3390/en13164183.
80. Z. Tong, J. Miao, S. Tong, and Y. Lu, "Early prediction of remaining useful life for Lithium-ion batteries based on a hybrid machine learning method," *Journal of Cleaner Production*, vol. 317, p. 128265, 2021/10/01/ 2021, doi: https://doi.org/10.1016/j.jclepro.2021.128265.
81. Q. Zhu, M. Xu, W. Liu, and M. Zheng, "A state of charge estimation method for lithium-ion batteries based on fractional order adaptive extended kalman filter," *Energy*, vol. 187, p. 115880, 2019/11/15/ 2019, doi: https://doi.org/10.1016/j.energy.2019.115880.
82. E. H. Karimi and K. B. McAuley, "A Bayesian Method for Estimating Parameters in Stochastic Differential," *IFAC-PapersOnLine*, vol. 48, no. 8, pp. 147-152, 2015/01/01/ 2015, doi: https://doi.org/10.1016/j.ifacol.2015.08.172.
83. M. Haridim and R. Zemach, "Stochastic Processes Approach in GPR Applications," *IEEE Transactions on Geoscience and Remote Sensing*, pp. 1-10, 2021, doi: 10.1109/TGRS.2021.3062427.
84. R. Janssen, "Discretization of the Wiener-process in difference-methods for stochastic differential equations," *Stochastic Processes and Their Applications*, vol. 18, no. 2, pp. 361-369, 1984/11/01/ 1984, doi: https://doi.org/10.1016/0304-4149(84)90306-5.
85. "Energy Storage Systems Market." https://www.alliedmarketresearch.com/advanced-energy-storage-systems-market (accessed.
86. S. R. Golroudbary, D. Calisaya-Azpilcueta, and A. Kraslawski, "The Life Cycle of Energy Consumption and Greenhouse Gas Emissions from Critical Minerals Recycling: Case of Lithium-ion Batteries," *Procedia CIRP*, vol. 80, pp. 316-321, 2019/01/01/ 2019, doi: https://doi.org/10.1016/j.procir.2019.01.003.

87. A. Azzuni and C. Breyer, "Energy security and energy storage technologies," *Energy Procedia,* vol. 155, pp. 237-258, 2018/11/01/ 2018, doi: https://doi.org/10.1016/j.egypro.2018.11.053.
88. A. Matallana *et al.*, "Power module electronics in HEV/EV applications: New trends in wide-bandgap semiconductor technologies and design aspects," *Renewable and Sustainable Energy Reviews,* vol. 113, p. 109264, 2019/10/01/ 2019, doi: https://doi.org/10.1016/j.rser.2019.109264.
89. B. K. Bose, "Chapter 1 - Energy, environment, and power electronics: a general introduction," in *Power Electronics and Motor Drives (Second Edition),* B. K. Bose, Ed.: Academic Press, 2021, pp. 1-57.
90. A. Sharma and S. Sharma, "Review of power electronics in vehicle-to-grid systems," *Journal of Energy Storage,* vol. 21, pp. 337-361, 2019/02/01/ 2019, doi: https://doi.org/10.1016/j.est.2018.11.022.
91. L. Shen, Q. Cheng, Y. Cheng, L. Wei, and Y. Wang, "Hierarchical control of DC micro-grid for photovoltaic EV charging station based on flywheel and battery energy storage system," *Electric Power Systems Research,* vol. 179, p. 106079, 2020/02/01/ 2020, doi: https://doi.org/10.1016/j.epsr.2019.106079.
92. X. Zhu, M. Xia, and H.-D. Chiang, "Coordinated sectional droop charging control for EV aggregator enhancing frequency stability of microgrid with high penetration of renewable energy sources," *Applied Energy,* vol. 210, pp. 936-943, 2018/01/15/ 2018, doi: https://doi.org/10.1016/j.apenergy.2017.07.087.
93. R. Gupta *et al.*, "Spatial analysis of distribution grid capacity and costs to enable massive deployment of PV, electric mobility and electric heating," *Applied Energy,* vol. 287, p. 116504, 2021/04/01/ 2021, doi: https://doi.org/10.1016/j.apenergy.2021.116504.
94. K. Turcheniuk, D. Bondarev, G. G. Amatucci, and G. Yushin, "Battery materials for low-cost electric transportation," *Materials Today,* vol. 42, pp. 57-72, 2021/01/01/ 2021, doi: https://doi.org/10.1016/j.mattod.2020.09.027.
95. F. Duffner, M. Wentker, M. Greenwood, and J. Leker, "Battery cost modeling: A review and directions for future research," *Renewable and Sustainable Energy Reviews,* vol. 127, p. 109872, 2020/07/01/ 2020, doi: https://doi.org/10.1016/j.rser.2020.109872.
96. M. H. S. M. Haram, J. W. Lee, G. Ramasamy, E. E. Ngu, S. P. Thiagarajah, and Y. H. Lee, "Feasibility of utilising second life EV batteries: Applications, lifespan, economics, environmental impact, assessment, and challenges," *Alexandria Engineering Journal,* vol. 60, no. 5, pp. 4517-4536, 2021/10/01/ 2021, doi: https://doi.org/10.1016/j.aej.2021.03.021.
97. Y. Bai, N. Muralidharan, Y.-K. Sun, S. Passerini, M. Stanley Whittingham, and I. Belharouak, "Energy and environmental aspects in recycling lithium-ion batteries: Concept of Battery Identity Global Passport," *Materials Today,* vol. 41, pp. 304-315, 2020/12/01/ 2020, doi: https://doi.org/10.1016/j.mattod.2020.09.001.

98. L. Barelli, G. Bidini, D. A. Ciupageanu, and D. Pelosi, "Integrating Hybrid Energy Storage System on a Wind Generator to enhance grid safety and stability: A Levelized Cost of Electricity analysis," *Journal of Energy Storage,* vol. 34, p. 102050, 2021/02/01/ 2021, doi: https://doi.org/10.1016/j.est.2020.102050.
99. A. Pfrang, A. Kriston, V. Ruiz, N. Lebedeva, and F. di Persio, "Chapter Eight - Safety of Rechargeable Energy Storage Systems with a focus on Li-ion Technology," in *Emerging Nanotechnologies in Rechargeable Energy Storage Systems*, L. M. Rodriguez-Martinez and N. Omar Eds. Boston: Elsevier, 2017, pp. 253-290.
100. Z. Xu, F. Zhang, M. Zhang, and P. Wang, "Energy storage development trends and key issues for future energy system modeling," *IOP Conference Series: Earth and Environmental Science,* vol. 526, p. 012114, 2020/07/08 2020, doi: 10.1088/1755-1315/526/1/012114.
101. T. Capuder, D. Miloš Sprčić, D. Zoričić, and H. Pandžić, "Review of challenges and assessment of electric vehicles integration policy goals: Integrated risk analysis approach," *International Journal of Electrical Power & Energy Systems,* vol. 119, p. 105894, 2020/07/01/ 2020, doi: https://doi.org/10.1016/j.ijepes.2020.105894.
102. S. Li, L. Tong, J. Xing, and Y. Zhou, "The Market for Electric Vehicles: Indirect Network Effects and Policy Design," *Journal of the Association of Environmental and Resource Economists,* vol. 4, no. 1, pp. 89-133, 2017/03/01 2017, doi: 10.1086/689702.

4

Thermal Management of the Li-Ion Batteries to Improve the Performance of the Electric Vehicles Applications

Hamidreza Behi[1,2*], Foad H. Gandoman[1,2], Danial Karimi[1,2], MD Sazzad Hosen[1,2], Mohammadreza Behi[3,4], Joris Jaguemont[1,2], Joeri Van Mierlo[1,2] and Maitane Berecibar[1,2]

[1]ETEC Department & MOBI Research Group, Vrije Universiteit Brussel, Brussel, Belgium
[2]Flanders Make, Heverlee, Belgium
[3]The University of Sydney, School of Chemical and Biomolecular Engineering, NSW, Australia
[4]Department of Energy Technology, KTH Royal Institute of Technology, Stockholm, Sweden

Abstract
In recent decades, clean energy has been introduced as an alternative energy source. In this respect, the automobile sector, one of the world's most prominent industries, has significantly contributed to the problem. Lithium-ion (Li-ion) batteries are among the most acceptable power sources for the automobile sector. Li-ion batteries benefit from high energy storage density, high cycle lifetime, and minimal self-discharge. Despite this, Li-ion batteries generate a significant amount of heat during the charge/discharge process due to chemical reactions and ohmic resistance. At the same time, Li-ion battery lifespan, efficiency, and safety are all influenced by temperature and temperature uniformity, which has become a significant factor affecting Li-ion battery performance. According to the studies, the suitable operating temperature for Li-ion batteries is slightly from about 25 to 40 °C, with a maximum temperature variance of less than 5 °C between cells and modules.

Consequently, designing a capable battery thermal management system (BTMS) has become a fundamental challenge in the automotive industry to

*Corresponding author: Hamidreza.behi@VUB.be

Pedram Asef, Sanjeevikumar Padmanaban, and Andrew Lapthorn (eds.) Modern Automotive Electrical Systems, (149–192) © 2023 Scrivener Publishing LLC

control and remove the generated heat by the cells from the safety and reliability evaluation. In addition, investigating the different thermal management systems (TMSs), their advantages, and disadvantages are critical in the automotive study. Therefore, a variation of active, passive, and hybrid cooling systems have been developed during the past decade to reach the heat-dissipation necessity for Li-ion batteries. The following chapter presents a comprehensive review of different cooling methods for battery thermal management at the cell, module, and pack levels.

The chapter is organised as follows. Section 4.2 represents the objective of the research. Section 4.3 presents electric vehicle (EV) trend. Section 4.4, presents the different TMSs of the Li-ion batteries. The lifetime performance of Li-ion batteries is described in Section 4.5. The fundamental aspects of safety and reliability evaluation of EVs are investigated in Section 4.6. Lastly, a relevant conclusion is drawn in Section 4.7.

***Keywords*:** Electric vehicle, lithium-ion battery, thermal management, active cooling system, passive cooling system, safety, reliability

Acronyms

Li-ion	Lithium-ion
EVs	Electric vehicles
TMS	Thermal management system
BEV	Battery electric vehicle
PHEV	Plug-in hybrid electric vehicle
BTMS	Battery thermal management system
PCM	Phase change material
HPCS	Heat pipe with copper sheets
SHCS	Sandwiched configuration of heat pipes air cooling system
LiBs	Li-ion battery technologies
SoH	State of health
DoD	Depth of discharge
SoC	State of charge
OEMs	Original equipment manufacturers
NMC	Nickel-manganese-cobalt
FEC	Full equivalent cycle
EoL	End of life

4.1 Introduction

Currently, the transportation industry focuses on clean energy vehicles because of global warming and environmental pollution. EVs and hybrid electric vehicles are presented as new technology for the transportation industry [1]. The traditional engine of vehicles is exchanged for an electric battery pack in EVs. In combustion engine vehicles, the energy needed to drive the vehicle is attained from the fossil fuels, whereas in EVs, energy comes from the battery, which is using power supply during the rest condition. This similar energy is used during the working condition of the vehicle. The application of EVs reduces a considerable amount of fossil fuel which is a consequence of the CO_2 emissions and control of global warming. Renewable energies comprising wind and solar energy can contribute to battery charging, leading to a decrease in environmental pollution.

The main component of EVs is the battery as a source of energy. Among all types of rechargeable batteries, Li-ion batteries have received more attention as a result of their unique features like low self-discharge, wide application, lightweight, high energy, high power density, and long life cycle [2, 3]. All types of EVs include an energy storage source like a Li-ion battery pack according to these features. EVs placement has been rising over the last decade, with the global standard of electric passenger cars passing 5 million in 2018, with an increase of 63% from the year 2017. Nevertheless, various challenges company with the development of EVs, for instance, heat generation of batteries, high investment cost, and safety. The battery is known as electrical energy storage, which releases the energy by electrochemical reactions. The battery's fast charging is an essential item to save the charging of the battery pack. In addition, fast discharging is an important feature of EVs for high acceleration. Though, fast charging/discharging generates a noticeable amount of heat inside the battery pack, leading to a thermal runaway, fire, and even explosion.

This heat generation also affects the temperature uniformity of the battery pack, which can cause uncertainty for the EVs. Many researchers proved that the ideal battery temperature is 25 °C-40 °C [4]. High temperature disturbs the charge/discharge efficiency, speeds up the capacity degradation, reduces the output power, and decreases the battery life. Since high temperature is not suitable, the low ambient condition is also destructive in terms of power capacity and lifespan. Therefore preserving the battery in the mentioned temperature range is an important issue. Consequently, designing an appropriate and capable thermal management system (TMS) to keep the battery temperature within the optimum range

Table 4.1 Literature study on battery thermal managment.

Authors	**Application**	**Active TMS**	**Passive TMS**	**Hybrid TMS**	**Safety**	**Reliability**	**Lifetime**
Sharma *et al.* [5]	TMS	✓	-	✓	-	-	-
Zhao *et al.* [6]	TMS	✓	-	✓	-	-	-
Murali *et al.* [7]	TMS	✓	✓	✓	-	-	-
Chacko & Charmer [8]	TMS	✓	-	-	-	-	-
Gandoman *et al.* [3]	Reliability and TMS	✓	✓	✓	-	✓	-
Sunden [9]	TMS	✓	✓	-	-	-	-
Klinger & Jagsch [10]	Thermal behaviour	✓	✓	✓	-	-	-
Thakur *et al.* [11]	Advance cooling system	✓	✓	✓	-	-	-
Lu *et al.* [12]	TMS	✓	✓	✓	-	-	-

and avoid thermal runaway is crucial for the performance of Li-ion batteries. Several studies have considered the effect of different cooling systems on Li-ion batteries. However, this chapter has considered a comprehensive study concerning different cooling strategies, lifetime, safety, and reliability of the Li-ion battery cells.

4.2 The Objective of the Research

Thermal management and cooling applications for Li-ion batteries are significant subjects directly influencing EVs' performance, safety, and reliability. Fundamentally the produced heat during the charging or discharging process returns to ohmic and electrochemical processes, which can be transferred by conduction through the different materials of the cell to the ambient. Many researchers have considered the batteries TMS. Table 4.1 shows several research studies on the application of battery TMS.

According to the literature review and the authors' knowledge, a few studies widely considered the TMS for Li-ion batteries. This chapter presents the vast expansion and arrangement of the different active, passive and hybrid cooling systems for EVs. Moreover, the lifetime, safety, and reliability of the Li-ion batteries have been considered in the chapter.

4.3 Electric Vehicles Trend

EV sales were around 1 and 2 million markets in 2017 and 2018, respectively [13]. Smaller automotive markets are leading the charge for EV, with a worldwide battery electric vehicle (BEV). Norway has the largest market share, with 56% of new cars sold in 2019 by electric or plug-in hybrid electric vehicle (PHEV). In second place is Iceland, with a share of 24.5%, followed by the Netherlands with a share of 15 %. China leads the global markets with a plug-in share of 5.2%, followed by the UK (3.2%), Germany (2.9%), France (2.8%), Canada (2.7%), and the United States (2.4%) [14].

Moreover, PHEVs and BEVs stock and market shares in the developed world, where these technologies have been mainly adopted [15]. Since 2010, battery density has grown, and EV costs have decreased. EVs are estimated to account for 20% of all vehicles on the road in China by 2040. This figure is likely to rise to 10% or more in the U.S. and Europe (the United Kingdom, Germany, and France) [15, 16]. According to a new report, more than 50 million EVs might be sold in China, Germany, and the United

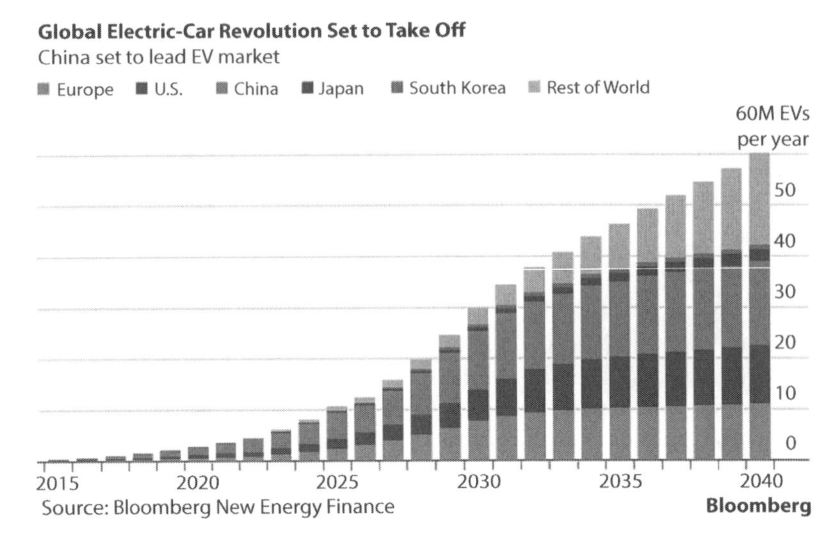

Figure 4.1 The global forecast EV yearly sales by 2040 [17].

States by 2040 [13]. Figure 4.1 shows the global forecast for annual EV sales by 2040.

4.4 Thermal Management of the Li-Ion Batteries

4.4.1 Internal Battery Thermal Management System

The internal battery thermal management system (BTMS) is the initial phase for revision of the battery, which progresses battery thermal performance by decreasing the internal resistance of the battery cells. The internal BTMS modification has been done by several researchers in anode/cathode material and electrode thickness. Lu *et al.* [18] and Zaho *et al.* [19] considered the effect of electrode thickness on the volume and weight of the battery cell. Sheu *et al.* studied the effect of particle size on electrode proficiency. They found that the smaller particle in the electrode results in lower thermal resistance. Some researchers are also done to improve the anode and cathode material with novel substances. However, it was found that modification of the anode and cathode material leads to degradation of the energy storage capacity of the cell, which is not recommended nowadays.

4.4.2 External Battery Thermal Management System

External BTMS refers to any system cooled by a battery cell/module by an external cooling medium. Generally, it is divided into active, passive, and hybrid cooling systems. Liquid and forced air cooling systems are examples of active cooling systems requiring an external energy source. However, passive battery cooling systems such as phase change material (PCM) and heat pipe are known as cooling systems without an external energy source. The combination of active and passive cooling systems can be classified into hybrid cooling systems. Figure 4.2 shows the detailed classification of the BTMS [3].

4.4.2.1 Active Cooling Systems

4.4.2.1.1 Air-Based Cooling System

Air cooling technology is based on convection heat transfer which can remove the heat away from the battery surface. Air cooling is a simple, famous, and easy cooling system [20]. This technology is used in former electric cars, like the Nissan Leaf and Toyota Prius. However, it is not recommended in stressful conditions and high power applications due to the air's low heat transfer coefficient. As EVs technology is being used more

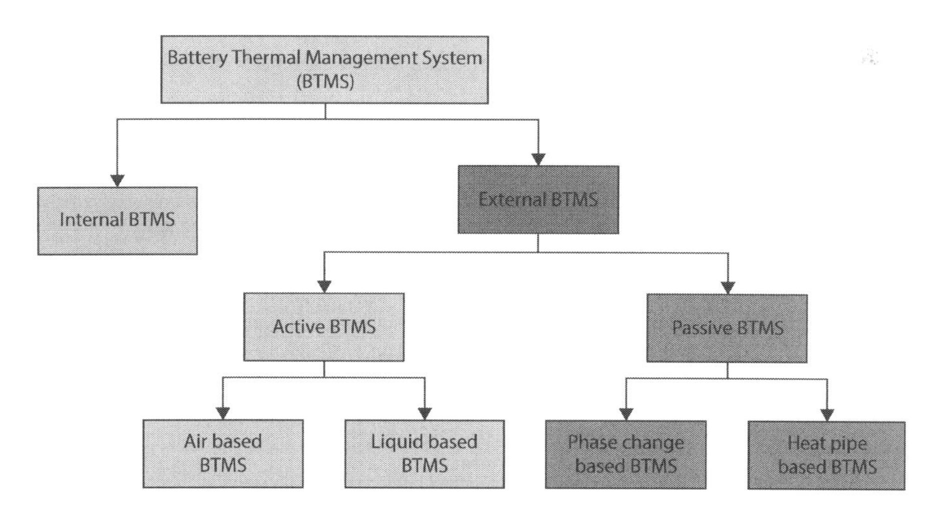

Figure 4.2 The detailed classification of the BTMS.

frequently, safety subjects have arisen with purely air cooling technology, particularly in hot climates. Several kinds of research have been done to increase the heat dissipation of the air cooling by modifying wind source layout, high thermal conductive metals, cooling condition, flow paths, and battery configuration [21].

Generating Joule losses and overheating of batteries are the reasons why a proper TMS is vital to avoid the lifetime degradation of batteries [22]. Also, the generated heat may result in non-uniform temperature evolution, and voltage unbalance of the battery [23]. Therefore, proposing a proper TMS including active and passive cooling systems would help batteries to operate in a safe operating area.

The typical active cooling TMS for batteries comprises air cooling and liquid cooling systems. The most common active cooling system is an air-cooled system due to its simple structure and low cost. Behi *et al.* [24] experimentally and numerically expand a hybrid air cooling method for LTO cells at the cell level. They investigated the effect of ambient temperature and inlet velocity on maximum cell temperature.

Shahid *et al.* [25] developed a new technique to improve air-cooling and temperature uniformity in a battery pack in which the problem of non-uniform cooling of cylindrical cells was addressed by adding an inlet plenum as the second inlet to the cylindrical battery pack with an axial airflow. Also, they studied mixing and turbulence in the airflow to enhance cooling and uniformity. Chen *et al.* [26] conducted an optimisation study for the structure of parallel air-cooled BTMS with U-type flow (BTMSU) and developed a simplified model named flow resistance network model for BTMSU. Also, they optimised the widths of the plenums and widths of the inlet and outlet based on the simplified model in which they improved the cooling efficiency of BTMSU remarkably. Lu *et al.* [27] conducted a numerical study for air cooling for Li-ion batteries packed with a staggered arrangement. They proved that a larger channel size lowers the maximum temperature and improves energy efficiency, and they recommended that the best configuration would be ten rows of cylindrical batteries along the airflow direction.

Chen *et al.* [28] used an approach for homogenising the airflow rates in which they included an adjustment coefficient to improve the battery cell spacing distribution of the parallel air-cooled battery pack. The cooling efficiency of the BTMS was significantly improved using this strategy. Jilte *et al.* [29] used PCM to modify an air cooling system, demonstrating that innovative cell-to-cell air cooling results in temperature rises of less than 5 °C and temperature homogeneity of the cells of less than 0.12 °C. As a consequence, the hybrid active-passive cooling system lowered the maximum module temperature. Peng *et al.* [30] established a heat generation model considering the inconsistent thermal property of cells in which they showed that an inlet/outlet on the same side of the pack ensures better thermal efficiency than a both-side placement. Also, they proved that inlet vent height has more impact on maximum temperature than the outlet vent height. Li *et al.* [31] improved heat dissipation capability using silica cooling plate coupled with a copper mesh-based air cooling system through experiments and simulations. Jiaqiang *et al.* [32] investigated the thermal characteristic of a battery module consisting of 60 cells using different air cooling strategies in which the inlet and outlet locations were changed to achieve the best cooling performance. Also, they utilised a baffle to enhance the airflow distribution in the module.

Chen *et al.* [33] improved the performance of air-cooled BTMS by designing the flow pattern and by changing the different positions of the inlet and the outlet. Akinlabi *et al.* [34] categorised tabular classification of recent studies on air cooling techniques and reported recent advancements to air cooling systems in the forms of hybrid TMSs. Also, they classified the parameter configuration optimisation techniques to improve the objectives of the air-cooled BTMS. Yu *et al.* [35] studied a forced-air cooling strategy with longitudinal airflow for a battery with $Li(Ni_xCo_yAl_z)O_2$ cathode material in which the proposed active cooling system with a flow velocity of $0.8\ m{\cdot}s^{-1}$ was recommended. Fan *et al.* [36] studied the influence of the discharge rate and air input temperature on the proposed cooling system was investigated using the thermal performance of aligned, staggered, and cross battery packs. Zhang *et al.* [37] improved the cooling performance of battery TMS by changing the number, position, angle, and height of spoilers. They also looked at the length and height of the manifolds, as well as the width of the cooling channel and spoilers in the cooling channel. Chen *et al.* [38] improved the thermal performance of the air-cooled BTMS by structural design, and the influence of the battery cell number was investigated. They constructed the symmetrical BTMS with uneven cell spacing to reduce the temperature difference by 43%.

Cheng *et al.* [39] designed a novel finned forced air cooling system and proposed a high-fidelity design optimisation method. They proved that using the new method, the average temperature of the module and pressure drop were reduced by 0.57% and 8.44%, respectively. Yang *et al.* [40] proposed an axial air-cooled BTMS using a novel radiator with the bionic surface structure for a battery module. In this method, when the inlet velocity increases, the cooling efficiency and power consumption of the module are increased as well. The main parameters to control the module temperature are radiator thickness and the bionic surface structure height

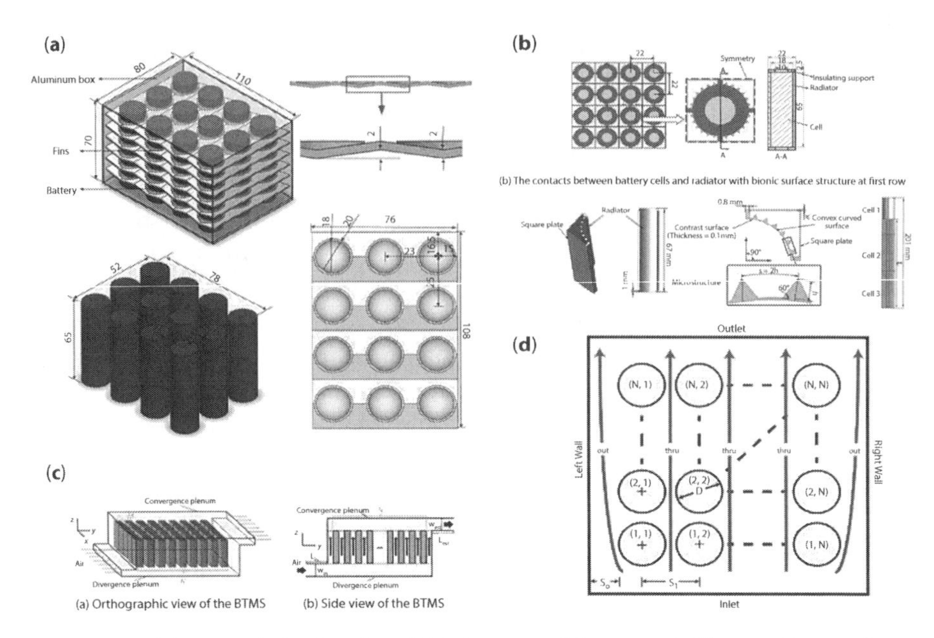

Figure 4.3 (a) Air-cooled BTMS [39] (b) Schematic of the air-cooled battery TMS with bionic surface structure [40] (c) Schematic of the parallel air-cooled BTMS [41] (d) Top view of flow paths for scalable cylindrical battery pack [42].

in which by using optimal values for these parameters, the best cooling performance of the air cooling system can be achieved. In Chen *et al.* [41], an air-cooled BTMS, a configuration optimisation of the battery pack was carried out, with an optimisation approach employed to optimise the battery spacings to minimise the battery pack's maximum temperature. Erb *et al.* [42] answered the question of "what size of a Li-ion cell is the best for an air-cooled battery pack?" by establishing an analytical framework to determine the min-cost cell size. They found that a non-optimal cell can double or triple the cost of a fan/blower. Behi *et al.* [43] experimentally and numerically studied the effect of the aluminium heat sink and air cooling on the LTO battery cell under an 8C discharging rate. Different configurations of air cooling systems for cylindrical/prismatic battery module/pack-based BTMS are shown in Figure 4.3.

4.4.2.1.2 Liquid-Based Cooling System

The low heat transfer coefficient of the air-cooled system made them an inefficient cooling system for high peak power applications. Therefore, a more reliable cooling system with a higher heat transfer coefficient is mandatory to control the maximum battery temperature in high-power applications. The liquid cooling system is classified as the most practical and promising cooling approach with a compact design. Liquid coolants have a higher heat transfer coefficient compared with air and consequently achieve more efficient cooling [44]. The liquid-based cooling systems benefit from their advantages like compact structure and ease of arrangement [45]. Typically, the liquid-based cooling system comprises the medium contact mechanism with the cell body, namely (i) direct contact method and (ii) indirect contact method.

This technology will bring the best performance to keep a battery pack in a safe temperature range and temperature uniformity. GM, Tesla, Jaguar, and BMW currently uses this cooling method.

Figure 4.4 shows the popularity of liquid-based cooling systems among all kinds on TMS. On the other hand, the risk of a short circuit would be higher in liquid-based systems due to leakage [46]. Moreover, additional components like the tube, coolant, and pump are needed for coolant circulation, which results in a heavier cooling system [47].

Zhou *et al.* [49] proposed a cooling strategy based on the half-helical duct for a cylindrical Li-ion battery module. They constructed a three-dimensional computational fluid dynamics model to reduce the module's maximum temperature and improve the temperature uniformity. Wang *et al.* [50] proposed an effective optimising route for liquid cooling technology for a cylindrical power battery module in which the influence of the contact angle is more significant than inlet velocity and channel number. Also, based on their system, the contact angle should be fixed at 70° degrees. Al-zareer *et al.* [51] investigated a novel approach for performance

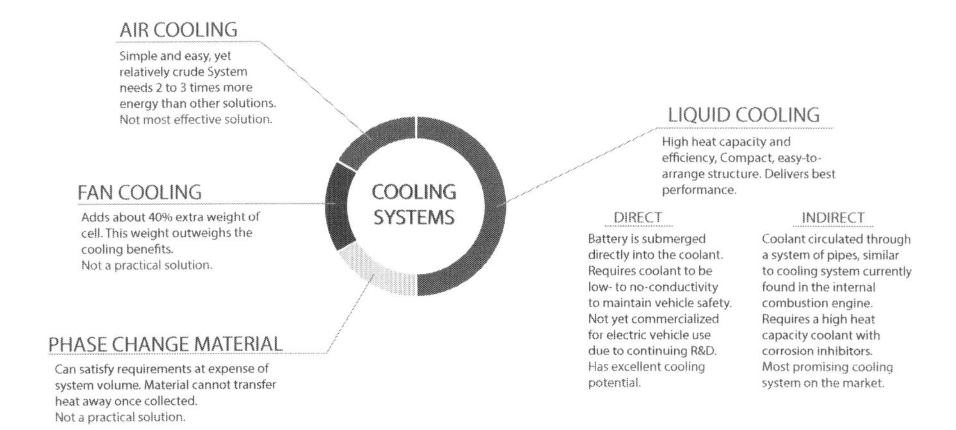

Figure 4.4 Comparison of different BTMS [48].

improvement of liquid to vapour-based TMS, which showed that battery arrangement affects the cooling performance of the TMS. They developed a boiling simulation model to predict the heat transfer rate of the boiling with an average error of 14.9%. Jiaqiang *et al.* [52] investigated the cooling effect of a liquid-cooled TMS in which an $L_{16}(44)$ orthogonal array was selected to design sixteen models. They showed that the number of tubes affects the average temperature of the cooling plate. Different configurations of liquid cooling systems for cylindrical/prismatic battery module/pack-based BTMS are shown in Figure 4.5.

Chen *et al.* [53] analysed a heat dissipation model with experiments to get thermal parameters. Also, they analysed the standard temperature deviation in thermodynamics for heat uniformity and considered the maximum pressure that has a huge impact on running costs. Yates *et al.* [54] investigated the effects of channel number, hole diameter, mass flow rate, and inlet locations in a liquid cooling system for Li-ion batteries. They proved that the mini channel-cooled cylinder design provides better performance compared with the channel-cooled heat sink to control the

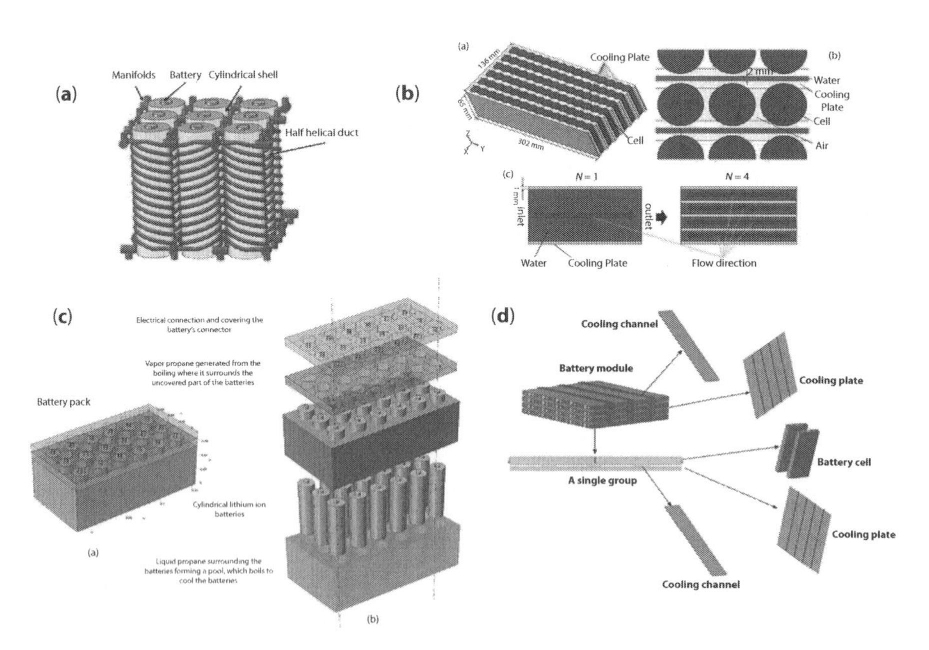

Figure 4.5 (a) Schematic diagram of battery module using half-helical duct [49] (b) Schematic diagram of the battery module, the interface between cells and LCPs, and the channels within the LCPs [50] (c) Schematic diagram of propane cooled battery pack of cylindrical batteries, where the pool covers the cylindrical batteries [51] (d) Schematic of liquid cooling battery module [52].

maximum temperature and temperature uniformity of the pack. Al-zareer *et al.* [55] conducted a comparative study to assess the recently developed categories of BTMS with a focus on liquid-to-vapour phase change cooling systems. Cao *et al.* [56] proposed a delayed cooling system for a large cylindrical battery module in which the hybrid PCM and liquid cooling heatsink were optimised. They found that higher inlet coolant temperature increases the battery discharge capacity. Besides, in the 4C discharge rate, delayed cooling enhanced the cylindrical battery module's temperature uniformity. The proposed TMS keeps the maximum battery module temperature below 55 °C with a temperature difference of less than 5 °C.

Lai *et al.* [57] proposed a compact and lightweight liquid-cooled BTMS and investigated the influences of the parameters on the performance of the thermally conductive structure. In such a robust TMS, pressure drop, temperature difference, and thermally conductive structure weight were reduced by 80%, 14%, and 46%, respectively. Zhu *et al.* [58] optimised a liquid-cooled system for a battery module in both axial and radial directions in which the maximum temperature of the battery module was reduced by 13 °C compared to the conventional design. To do so, sensitivity and single-factor analysis were carried out to identify the impacts of influence parameters. They showed that the thermal column diameter in the sensitivity analysis has the highest effect on optimisation objectives. Also, the maximum temperature and pressure drop were minimised in the multi-objective optimisation to obtain a lower temperature difference. Yang *et al.* [59] proposed a composite TMS integrated with a mini-channel liquid and air cooling system. They constructed a three-dimensional computational fluid dynamics model for the module to improve the temperature uniformity and maximum temperature reduction. They proved that by the integration of the air cooling with the liquid cooling system, the battery temperature could be controlled in the desired range. Huang *et al.* [60] numerically studied the instantaneous characteristic of a battery module. They showed that increasing the inflow rate results in module temperature reduction. Moreover, a hysteresis effect can be seen when the inlet flow rate changes.

Tang *et al.* [61] designed a novel liquid cooling BTMS composed of heat-conducting blocks with gradient contact surface angles. They obtained the optimum gradient contact surface angle of the blocks under preheating conditions that can guarantee the temperature uniformity of the battery module. Wu *et al.* [62] optimised a battery module based on hybrid liquid cooling with high latent heat PCM in which the maximum battery temperature and temperature difference are lower than a single liquid cooling system. They conducted sensitivity and single-factor analysis to identify influence parameters' effects and reduce the temperature difference. Yuan *et al.* [63] proposed a liquid cooling TMS and used a heat pipe to solve the issue of coolant leakage in the existing liquid cooling systems. The thermal performance of the proposed system was stable in continuous charging/discharging cycles, which ensures the long-term operation requirements of the battery. Wang *et al.* [64] designed a modular liquid-cooled BTMS for cylindrical Li-ion batteries. An experimental test bench identified the physical parameters of the cells, and the parameters were used as an input for numerical simulation studies. They proved that parallel cooling enhances thermal equilibrium behaviour. However, increasing the coolant flow rate would be a limit to further improving the cooling performance of the proposed BTMS. Xu *et al.* [65] proposed a composite silica gel plate to enhance the cooling efficiency and used an expanded graphite and copper foam to improve the thermal conductivity of the composite silica gel plate. The results showed that the composite silica gel plate transfers the generated heat to cooling tubes timely and promptly; in the liquid cooling TMS, using the composite gel has a better cooling performance.

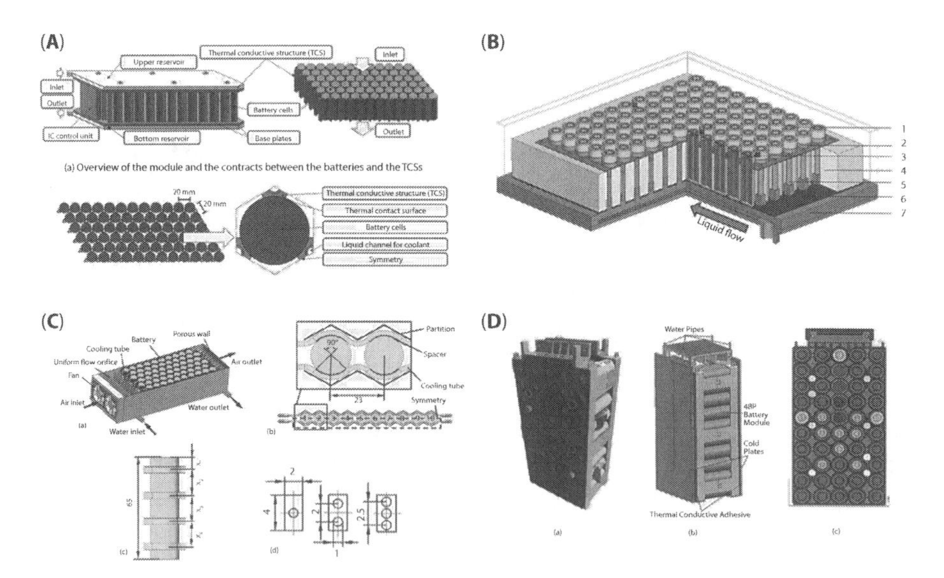

Figure 4.6 (A) Schematic of the liquid-cooled battery management module [57] (B) Geometry model of the battery module: 1-battery, 2-heat spreader plate, 3-the first temperature monitor point along the centerline, 4-air, 5-thermal column, 6-insulation plate, 7-cold plate [58] (C) Schematic diagram of battery module [59] (D) Experiment test object and internal temperature measurement point distribution; (a) Battery module, (b) Arrangement of cold plates relative to the battery module, (c) Thermocouple distribution points [60].

Several structures of liquid cooling systems for cylindrical/prismatic battery module/pack-based BTMS are also shown in Figure 4.6.

4.4.2.2 Passive Cooling Systems

4.4.2.2.1 Phase Change Based Cooling System

PCMs are materials that absorb or release huge amounts of energy when they go through a phase change in their physical state, from the solid to liquid and vice versa [66]. Phase change can happen in a heating or a cooling process immediately when the material reaches its phase change temperature. The temperature of the PCM remains constant during the phase change process, which is an essential feature of the PCM. This feature can be used to develop the thermal performance of different cooling methods in which the PCMs are applied [67]. Figure 4.7 shows the heat transfer mechanism in PCMs. PCMs are classified into three major groups: organic materials, inorganic materials, and eutectics [68]. Important characteristics like chemical stability, low vapour pressure during melting, and total recyclability have changed the paraffin wax into suitable candidates in cooling and energy storage applications. The advantages of uniform cooling, simple layout, and no power consumption increase the PCM usage for cooling applications.

However, most PCMs suffer from low thermal conductivity, a limited temperature range of phase change, and the leaking problem during the melting process. Therefore, many studies have been done to compensate for the thermal conductivity of PCMs using nanomaterials, fins, meshes, graphite, and heat pipes [70–72].

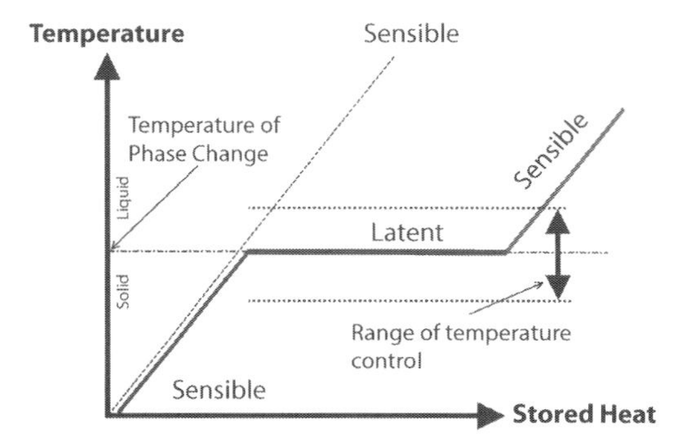

Figure 4.7 PCM heat storage as a latent heat [69].

Before-mentioned active cooling systems consume energy, but in contrast, passive cooling systems such as PCM and heat pipe do not consume energy or need low energy to operate [73, 74]. PCM is a material that absorbs and releases the generated heat of batteries during the phase transition process [75]. The first idea of using PCM in batteries was proposed by Al-Hallaj and Selman [76]. They proposed a novel TMS that incorporates PCM for EV applications and used finite-element software to simulate the temperature distribution of a battery module. Gradually, other researchers started investigating PCMs for battery TMS. Wu *et al.* [77] proposed a novel BTMS with flexible PCM with a technically simple and facile structure. As a result, the maximum temperature of the system using the proposed PCM-based TMS is 43.4 °C during the 2.5C discharge rate, which is 28.8 °C lower than no PCM. Behi *et al.* [78] investigated the feasibility of using PCMs in two different configurations for cold and heat storage using computational fluid dynamics. Also, they used PCM charging/discharging power and the solidification/melting process of the PCM in an integrated storage compartment. Behi *et al.* [79] investigated a PCM-based TMS with a melting domain between 25 °C and 32 °C for Li-ion battery cells under fast discharging. Yang *et al.* [80] numerically and experimentally analysed a paraffin PCM for cylindrical cells in the isothermal temperature. Several structures of phase change-based cooling systems for cylindrical/prismatic battery module/pack are shown in Figure 4.8.

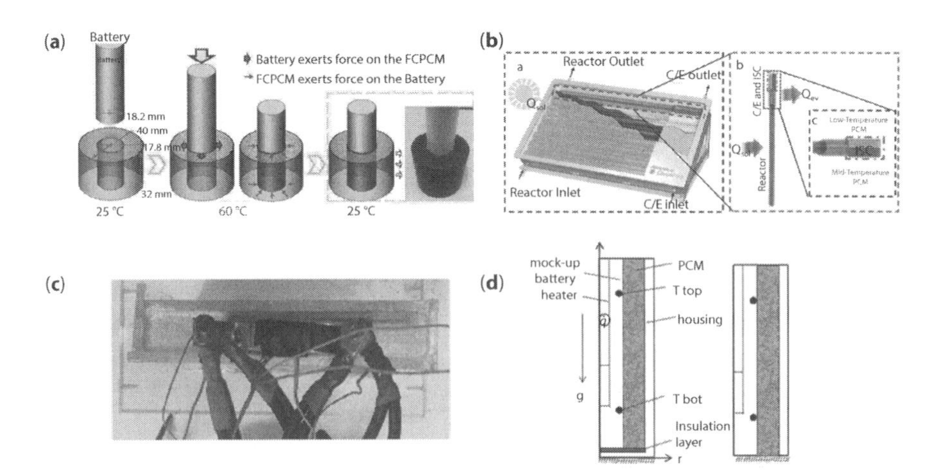

Figure 4.8 (a) Schematic diagram of the PCM-based TMS for the battery [77] (b) - (a) The solar thermal collector, (b) The sorption module compartments components (c) The integrated storage compartment [89] (c) using PCM for fast discharging of LTO cell [81] (d) Schematic of the baseline case and the case with full adiabatic base-plate [80].

However, almost all PCMs suffer from low thermal conductivity, which results in large amounts of heat accumulation in a harsh operating environment. Hence, the next target was the investigation of this issue by using PCM in combination with different high thermally conductive materials. Nemati and Pircheraghi [82] fabricated a PCM with green fatty acid and recycled silica nanoparticles to increase thermal conductivity in which silica nanoparticles were recycled from waste lead-acid battery separators. The new material had a significant enhancement of thermal conductivity because of the PbO inclusion. Also, the new form-stable PCM had an increased mechanical strength because of the sintering of high-density polyethylene powder with silica nanoparticles. Huang *et al.* [83] used a thermally induced flexible and form-stable PCM with a great battery thermal management impact. They found that styrene–butadiene–styrene exhibit an excellent impact on the structural stability of PCM composite. Also, they used aluminium nitride to enhance the thermal conductivity of the PCM. Li *et al.* [84] applied a microencapsulated PCM suspension to analyse the effects of the adjustable parameters of the microencapsulated PCM on the performance of the TMS. Zhang *et al.* [85] proposed a flame retardant from a stable PCM composite to enhance thermal properties. Two types of flame retardants were investigated, including ammonium polyphosphate and red phosphorus. They found that PCM composite with flame retardant additives has heat endurance and can increase thermal conductivity. Lv *et al.* [86] designed serpentine composite PCM plates to enhance heat dissipation performance by providing airflow channels and a large surface area. They showed that the new cooling structure saves 70% of the composite PCM in the battery module in which the energy density of the module is enhanced by 13.8 $Wh.kg^{-1}$.

Moreover, to make a robust cooling system for EVs, some researchers combine an active cooling system with the passive PCM-based TMS. Li *et al.* [87] investigated the liquid cooling with impregnated PCM and incorporated the interfacial gaps to analyse the numerical simulation. They showed that differential melting in PCM composite results in a higher temperature difference. Besides, maximum battery temperature and temperature differences were reduced by a double-sided cooling scheme. Lei *et al.* [88] proposed a hybrid TMS using PCM, heat pipe, and spray cooling for cooling at moderate and high temperatures. They showed that thermal storage by PCM protects the battery from low-temperature damage. Also, heating and cooling for the battery in various thermal conditions were rendered in the design. Cao *et al.* [89] presented a cooling system by ultrathin cold plate and nano-emulsion with no subcooling for Li-ion battery under high-rate discharge. Based on their work, lower maximum temperature and temperature differences were obtained when the batteries were cooled by nano-emulsion. Kiani *et al.* [90] introduced a new design for cooling Li-ion batteries using Al_2O_3/AgO/CuO nanofluids and PCM. They controlled the surface temperature of the battery in the hybrid TMS less than 5 °C. Also, it was shown that a heat sink with fins could remarkably prolong the working time of the battery. Different configurations of PCM-assisted high conductive materials for cylindrical/prismatic battery module/pack-based are shown in Figure 4.9.

Molaeimanesh *et al.* [91] studied the configuration performance of hybrid TMS comprising PCM and water-cooling channels for Li-ion batteries. They found that the parallel/series configuration exhibits excellent performance for the long-time operation of batteries, but for short period usages, one of the series configurations had better performance. Choudhari *et al.* [92] numerically investigated a BTMS using PCM at different current rates and analysed different fin structures by studying different fin

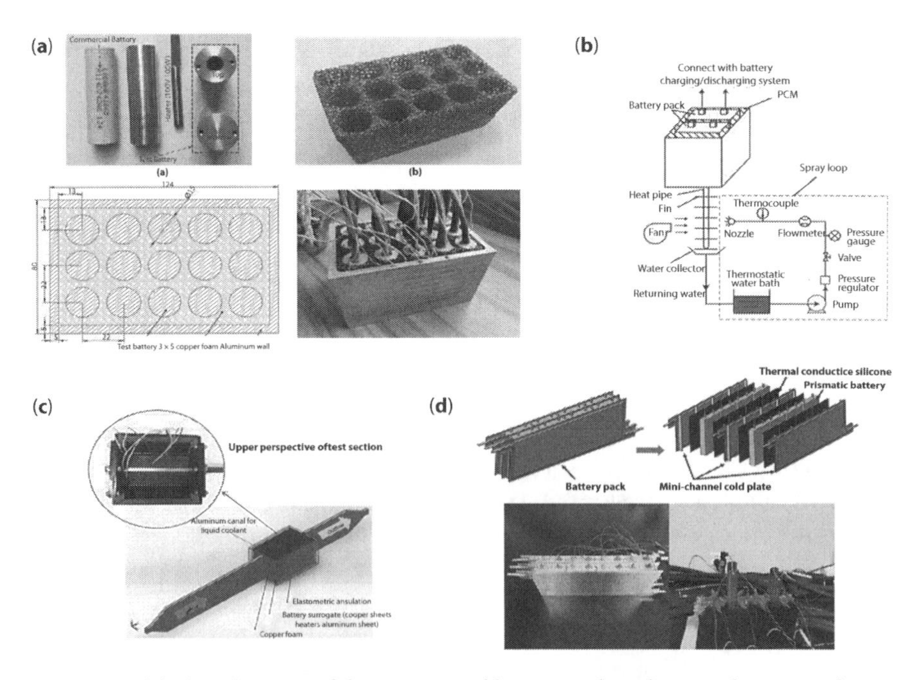

Figure 4.9 (a) The schematic of the commercial battery and test battery, the copper foam matrix, and the test battery module with composite PCM [87] (b) The schematic of the hybrid TMS based on PCM and spray cooling [88] (c) The schematic and photograph of the battery pack [89] (d) Schematic of the test bench [90].

shapes, fin numbers, and outer heat transfer coefficients. They proposed an optimal design for the PCM-based TMS. Wencan *et al.* [93] suggested a novel BTMS based on PCM and a liquid cooling system in which the PCM behaved as a heat buffer to avoid thermal runaway. Liu *et al.* [94] designed a battery cooling system based on the coupling of liquid cooling and a high-performance PW/EG/HDPE/nano-Ag PCM composite. In their structure, the liquid cooling system removed the heat absorbed by the PCM. This method is suitable for high charge/discharge current rates. Bamdezh *et al.* [95] proposed a hybrid TMS with cooling channels and PCM/Aluminum foam composite, that the TMS act for PCM recovery and cell cooling. Moreover, increasing tangential conductivity and axial thermal conductivity would reduce the temperature of the cell. Zhang [96] proposed an aluminium nitride-enhanced PCM composite to increase the thermal conductivity, stability, volume resistivity, and mechanical strength of composite materials. Joshy *et al.* [97] investigated the impact of vibration on PCM-based BTMS and characterised the temperature profiles under vibration utilising four distinct regimes. They found that battery surface temperature increases with increasing vibration frequencies and increasing discharge rates. Wang *et al.* [98] investigated a PCM OP28E nano-emulsion-based liquid cooling for a Li-ion battery pack. It was proved that by increasing OP28E concentration, the maximum battery temperature and temperature difference of the battery pack would be decreased. Similarly, liquid cooling TMS based on 10% weight of OP28E nano-emulsion had better performance than pure water.

4.4.2.2.2 Heat Pipe-Based Cooling System

The heat pipe is known as a two-phase heat transfer device that can transfer the heat at a high rate with a very low-temperature difference along the heat pipe length [99]. Usually, it consists of a container, a wick structure, and a working fluid. The container of a heat pipe is a sealed envelope that covers the wick and working fluid and can be made of metals, glass, or even plastic. Wick is a porous layer that is stuck to the inner surface of the container. This structure using capillary force, allows the working fluid to return from the condenser section to the evaporator section of the heat pipe. Wick structure is necessary for applications of zero gravity and where the evaporator is located at a higher level than the condenser. Screen mesh, sintered powder, and grooved are the most common wicks structures.

Figure 4.10 shows the heat transfer mechanism of the heat pipe. The working fluid also plays a vital role in the thermal performance of the

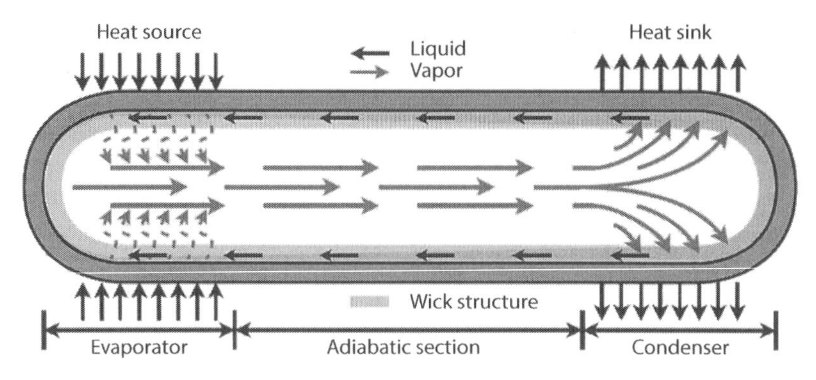

Figure 4.10 Heat transfer mechanism of the heat pipe [69].

heat pipe. Water, methanol, and ammonia are the common working fluid that is selected based on the Merit number [3].

Heat pipes have many advantages:lexible geometry, low cost, no maintenance, high heat transfer performance, and high efficiency [100]. They act as a superconductor, and their thermal conductivity reaches up to 90 times more than a copper fin with the same dimension. However, a secondary cooling system should cool the heat pipe condenser section. Moreover, they suffer from complex design, gravity, liquid, and vapour flow pressure drops. Therefore, many researchers studied the cooling effect of the heat pipe in different designs for BTMS.

The heat pipe as a superconductor is used in different thermal management applications for electronics and battery cooling. The heat pipe cooling technology can be combined with the liquid cooling, air cooling, and phase change cooling method. Li *et al.* [101] explored experimentally and numerically the impact of a liquid cold plate and heat pipe on a LiFePO4 battery pack. They studied the impact of heat pipes, coolant volumetric flow rate, and ambient temperature on the battery pack's maximum temperature. Behi *et al.* [102] considered the effect of liquid cooling and heat pipes on a single cell and the module at a high current. They found a 29.9% and 32.6% temperature reduction for a battery module using a liquid cooling system and LCHP, respectively. Liang *et al.* [103] studied the effect of heat pipe-based TMS under different coolant temperatures for a serially connected battery module. They investigated the effect of dynamics temperature, local current density, and voltage in the cooling process with different coolant temperatures. Gan *et al.* [104] developed a heat pipe-based liquid-cooled system for the thermal management of a cylindrical battery module. They studied the effect of different discharge rates, coolant flow rates, and inlet coolant temperatures. Behi *et al.* [105] introduced a novel thermal management method for a cylindrical battery module employing air cooling and a heat pipe. To increase the cooling performance of the module, they created a heat pipe with copper sheets (HPCS). They could regulate the module temperature for the cooling strategy utilising forced-air cooling, heat pipe, and HPCS at 42.4 °C, 37.5 °C, and 37.1 °C, respectively, lowering the module temperature by 34.5%, 42.1%, and 42.7%. Different configurations of heat pipe-based cooling systems for cylindrical/prismatic battery module/pack are shown in Figure 4.11.

Behi *et al.* [106] experimentally and numerically designed a sandwiched configuration of heat pipes air cooling system (SHCS) for a prismatic battery cell in high current applications. They investigated the feasibility of the cooling system in four different scenarios. They discovered that for the cooling scenario by natural convection, forced convection for SHCS, and forced convection for cell and SHCS, the maximum cell temperature equipped with SCHS decreased by up to 13.7 %, 31.6 %, and 33.4 %, respectively. Putra *et al.* [107] experimentally considered the effect of PCM-assisted heat pipe for thermal management of the EVs. They examined the cooling performance of the beeswax and RT44 for the battery simulator. Zhang *et al.* [108] designed heat pipe-assisted TMS based on

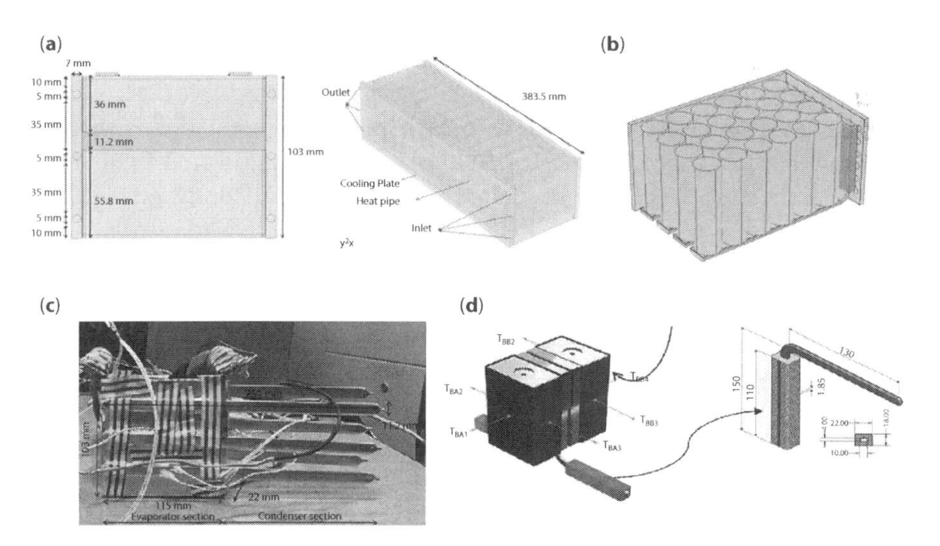

Figure 4.11 (a) The schematic of the commercial battery module equipped with liquid cooling and heat pipes [102] (b) The schematic of the cylindrical battery module equipped with HPCS [105] (c) The picture of the cell equipped with SHCS [106] (d) Schematic of the test bench [107].

PCM in which an additional fan and PCM were utilised to increase heat dissipation rate. They experimentally considered the cooling capacity of the TMS on the battery pack under discharging rates of 1C, 3C, 4C, and 5C. Huang *et al.* [109] experimentally studied the effect of the heat pipe-assisted PCM for a cylindrical battery module. They proved that heat pipe played an important role in cooling and temperature uniformly of the battery module.

4.5 Lifetime Performance of Li-Ion Batteries

The lifetime of Li-ion battery technologies (LiBs) is one of the key research topics to progress the expansion of EVs minimising range anxiety. The mileage of EVs solely depends on the battery energy capabilities which are directly linked to the state of health (SoH) of the battery. The SoH is a fractional term comparing the actual capacity to the nominal value often expressed in percentile form. In automotive applications, a battery needs to be replaced when the SoH of the battery system falls below 80% of its nominal content and should be replaced [110].

The aging of a LiB cell starts right from the moment when the final formation step is completed. Irrespective of the load or no-load (storage) situation, batteries tend to lose their capacity over time. When an external signal is employed to charge and/or discharge the battery, the cyclable Li-ions get mobile and start degrading depending on multiple interrelated factors [111]. It has been proved by researchers that aging happens with or without load conditions. Thus, the lifetime is merely connected to the battery chemistry and its capability to withstand operating conditions [112].

A battery has a highly nonlinear and complex electrochemical system that has an interdependent degradation characteristic. Moreover, this degradation scenario varies from technology to technology, but the basic mechanisms in LiBs are similar [113]. The aging of a battery cell or the SoH is typically expressed in terms of battery capacity fade, which happens due to several operating conditions [114]. These stressful conditions can be divided into two categories to represent the degradation forms of cycle life and calendar life. While the cycling depth of discharge (DoD), middle state of charge (SoC) region, number of cycles, temperature, and charge-discharge current rates (C-rate) affect the cycle life, the calendar life degradation factors include storage temperature, SoC, and duration [115]. These impact factors contribute to the cyclable lithium loss in the battery cell at different scales and accelerate the capacity fade if more than one parameter is deployed, especially outside the safe operating region [116].

However, researchers have succeeded in modelling the capacity degradation of different cell technologies by estimation methodologies which have benefited the original equipment manufacturers (OEMs) in improving the battery lifetime performance [117, 118].

4.5.1 Why Do Batteries Age?

The battery degradation is caused by several physical and chemical mechanisms related to the crucial cell components. These fundamental constituents include electrodes (anode and cathode), electrolytes, separators, and current collectors [119]. The degree of degradation is related to the different aging mechanisms and interdependencies. Figure 4.12 shows the common aging mechanisms in Li-ion battery cells.

Under load and no-load situations, the interrelated degradation factors get accelerated, and over time, the total aging takes control of the performance output. The demeaning of the battery performance may typically be expressed in two types of fading results commonly known as the capacity fade and internal resistance growth. Both of these resulting forms are caused by interconnected aging mechanisms and related to reduced energy and power capabilities, respectively.

4.5.2 Characterisation Techniques of Aging

A detailed and extensive lifetime characterisation needs to be done to investigate the performance characteristics of different battery cells under varieties of load conditions. Researchers have studied and considered

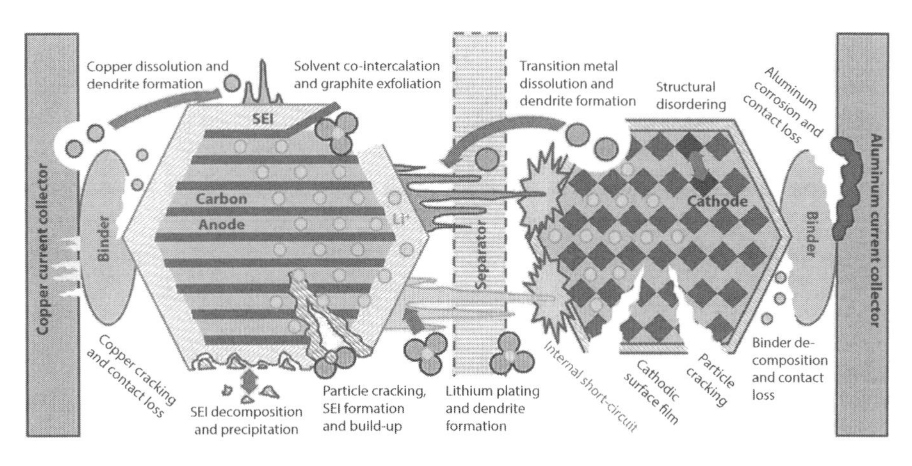

Figure 4.12 Common aging mechanisms in Li-ion battery cells.

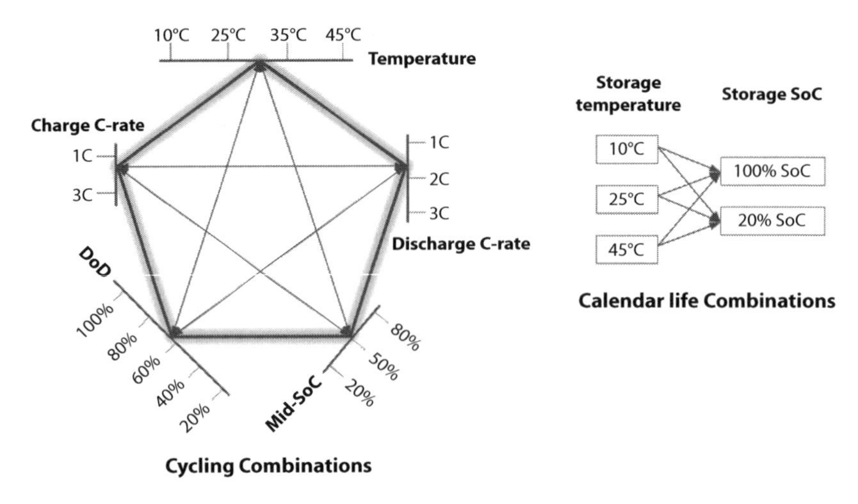

Figure 4.13 Cell aging characterisation matrix designed for an NMC cell.

many crucial stress factors that reduce battery performance to a different extent [118, 120]. These factors can be categorised by cycling (load) and calendar life (no-load) aging types.

- Cycle life factors: DoD, temperature, middle state of charge region (mid-SoC), C-rates, Ampere-hour throughput, and cycle number.
- Calendar life factors: Storage temperature, SoC, and duration.

While studying the battery behaviour under different impact factors, the selection of relevant test conditions is very important. The boundary of the investigated conditions would define a validity range of the characterised data facilitating the model development process. The test matrix is often designed carefully, considering all the relevant aspects of the long-term aging factors. Figure 4.13 displays the combination matrix of several connected aging parameters designed and performed for a nickel-manganese-cobalt (NMC) cell. The ranges of the used condition parameters are just a subset of a complete scenario; however, the selection of an optimised condition table may establish a strong base to generate comparable aging results.

4.5.3 Lifetime Tests Protocols of the Li-Ion Batteries

The test condition variants are generally employed to the Li-ion cells following indifferent test methodology. A typical test flow includes several rounds of battery cycling and/or calendar life storage tests together with reference performance tests to identify the battery SoH. Authors have described such a test flow chart in Figure 4.14 in which several stages of battery characterisation are performed during a year's long battery aging study. The full equivalent cycle (FEC) counter is used to track the cycles which are based on nominal charge ampere-hour throughput. This way shallow or full DoD cycling results could be compared on a single scale identifying the underlying reasons for the studied cell's aging mechanism. On the other hand, the check-up procedures include regular capacity and internal resistance tests, etc. to calculate the battery health. In the automotive industry, battery cell is typically marked as dead or at the end of life (EoL) when its capacity drops below 20% of its nominal value and/or the internal resistance increases over 100% of the nominal value [110].

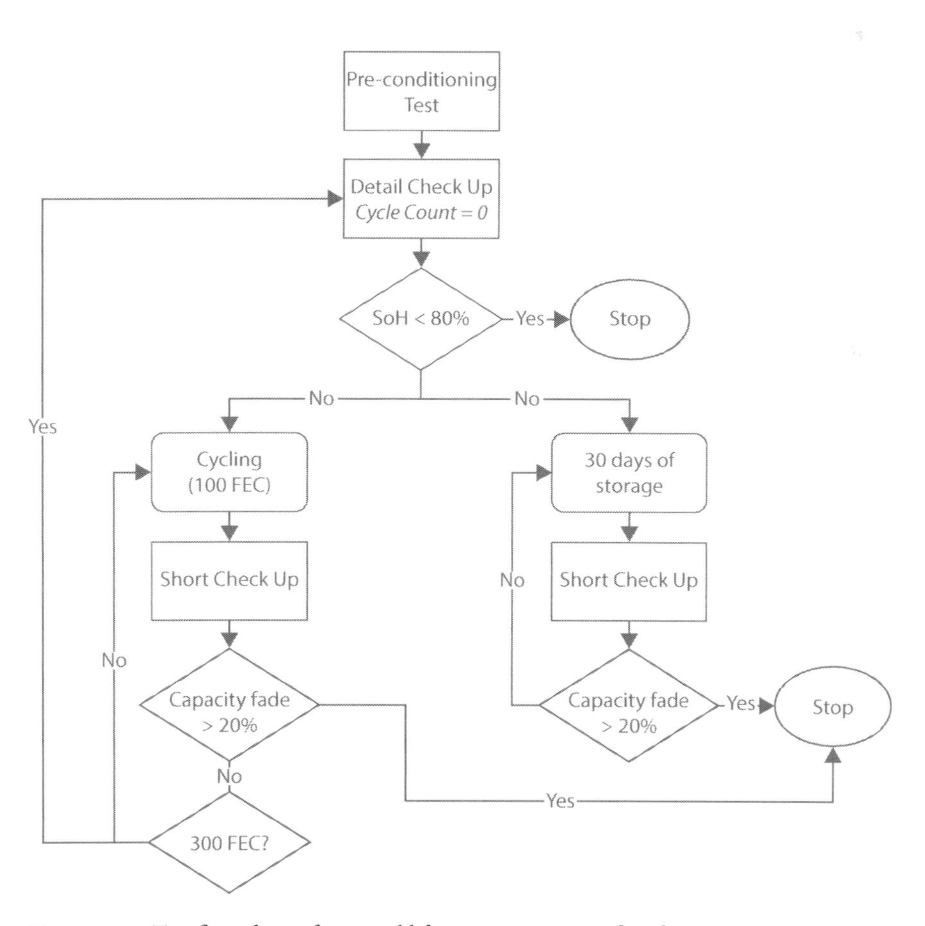

Figure 4.14 Test flow chart of a typical lifetime investigation [112].

4.5.4 Lifetime Results of Different Li-Ion Technologies

The importance of a lifetime investigation lies in the performance analysis that clarifies the aging scenarios. Depending on the stress factors, the cell capacity can drop following a certain pathway and the internal resistance can grow at the same time. Degradation results based on shallow or full cycling display different aging mechanisms (at different magnitudes) depending on the stress factors like temperature, C-rates, etc [120]. Figure 4.15 shows such capacity fade scenarios where the battery capacity fade is relatively expressed. It clearly shows the temperature dependency of the performed cycling for this cell type, but the difference is not significant at low temperature.

Such scale of aging performance is indifferent from cell to cell due to their differences in technology, type, and formation. Even the results may differ if the test procedures are slightly tuned. Thus, it is important to standardise and design the test methodology as long-term lifetime procedures may mislead to wrong interpretation. Hence, battery aging based on actual drive profiles is more realistic to investigate the degradation scenario. The lifetime investigation of a battery cell is a challenging and critical task as several dependence factors are to be considered. However, a well-designed and performed aging study may clarify the sensitive parameters responsible for the drop in battery performance. Understanding the aging mechanism

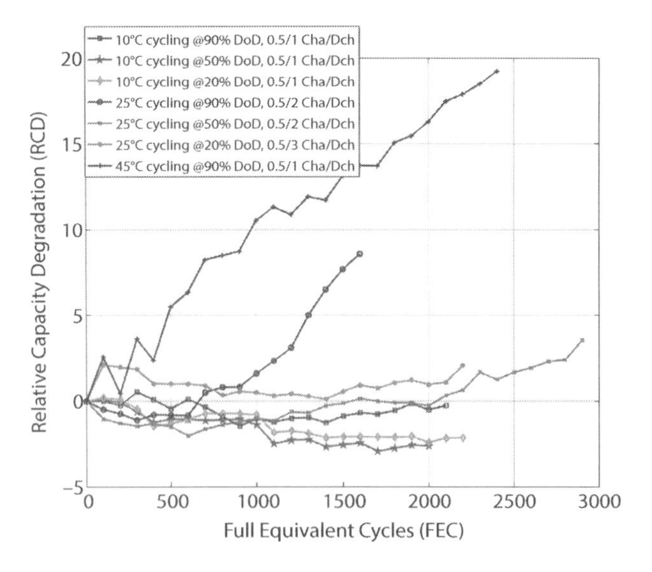

Figure 4.15 Battery relative capacity degradation comparison for a 20 Ah NMC cell when cycled at different DoD and with charge-discharge rates [112].

would further help the manufacturers and end-users optimise their usage to improve the performance prolonging the battery life.

4.6 Basic Aspects of Safety and Reliability Evaluation of EVs

Internal system reliability assessment is a major obstacle to the widespread use of EVs in present transportation networks. These are the most important issues to keep in mind while evaluating the reliability of EVs. [121–123]:

- The functionality of power components work.
- Detecting component problems in EVs.
- The failure sequences are inferred.
- Using a model to demonstrate the flaws.
- Choosing a strategy for assessing the dependability of EVs.

It is important to include breakdowns in electrical components when assessing the dependability of EVs. Next, the dependability of the electric components is assessed by all of the available methodologies. Table 4.2 shows the most typical electrical component failures in EVs.

Table 4.2 Classification of the main electrical components' failures in EVs [1].

Battery system failures [124–127]	Battery voltage failures	Low voltage failures Overvoltage failures High-voltage insulation failures High-voltage loop failures
	Battery current failures	Short circuit failures
	Battery temperature failures	Low-temperature failures Over-temperature failures
	Battery state of charge failures	Pre-charge failures Overcharge failures Over-discharge failures

(*Continued*)

Table 4.2 Classification of the main electrical components' failures in EVs [143]. (*Continued*)

Electric Motor failures [128–130]	Electrical failures	Rotor-related failures	Broken rotor bar Cracked rotor end rings Shorted rotor field windings
		Stator-related failures	The abnormal connection of stator windings The open or short circuit of stator windings
	Mechanical failures	Bearing failures	Outer bearing race defect Inner bearing race defect Ball defect Train defect
		Eccentricity-related failures	Bend shaft Static air-gap irregularities Dynamic air-gap irregularities
Power electronics failures [131–133]	Open switch failures	Current vector trajectory The 3-phase current mean value Voltage	
	Short switch failures	Phase current Device current Gate voltage	

4.6.1 Concept Reliability Analysis of Battery Pack from Thermal Aspects

Each Li-ion battery cell's electrochemical, thermal, and mechanical parameters have a significant relationship with reliability evaluation. However, while evaluating the battery pack's dependability, it's important to examine the reliability of individual cells and the reliability of other components in the battery pack, such as the TMS and the battery housing.

4.6.2 Reliability Assessment of the Li-Ion Battery at High and Low Temperatures

In this section for better understanding the concept of reliability, a dedicated test methodology is developed based on the NMC 3Ah cell to investigate the reliability of the Li-ion batteries at high and low temperatures under the same charge and discharge test protocols. Moreover, capacity fade has been considered a reliability indicator. The characteristic of the cylindrical NMC-based cell is reported in Table 4.3.

Table 4.3 Nominal specification of NMC 3Ah cell [134].

Item	Condition	Specification
Capacity	Std. charge/discharge	Nominal 3000 mAh
Nominal Voltage	Average for Std. discharge	3.60V
Standard Charge	Constant current	1500mA
	Constant voltage	4.2V
	End condition (Cut off)	50mA
Standard Discharge	Constant current End voltage (Cut off)	600mA 2.0V
Fast Discharge	Constant current End voltage (Cut off)	10000mA, 20000mA 2.0V
Discharge Current	For continuous discharge	20000mA
Operating Temperature (Cell Surface Temperature)	Charge Discharge	-5 ~ 50°C -20 ~ 75°C
Storage Temperature	Month Three months One year	-20 ~ 60°C -20 ~ 45°C -20 ~ 20°C

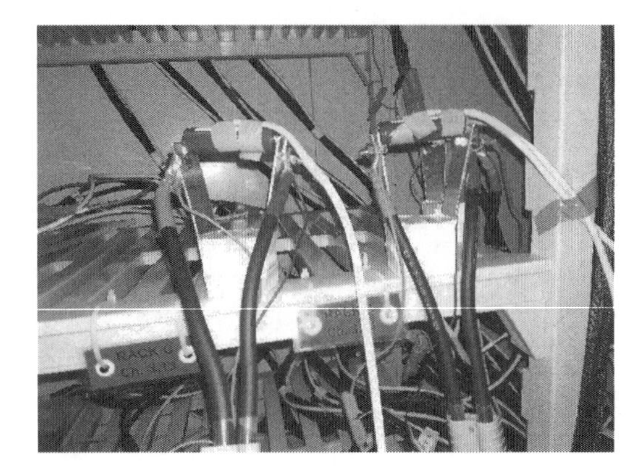

Figure 4.16 Li-ion batteries under the test at the 10 °C and 45 °C in the climate chambers.

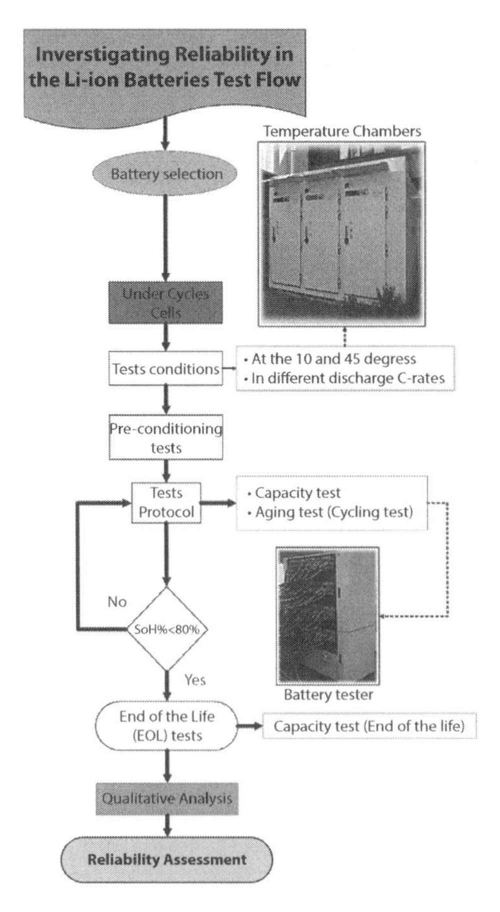

Figure 4.17 Investigation reliability in Li-Ion battery test flow.

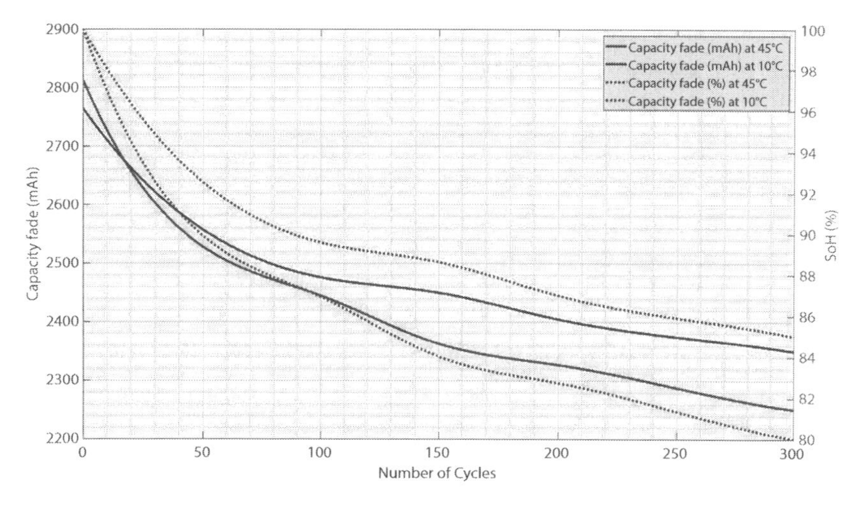

Figure 4.18 Capacity fade variation of the battery at 10 °C and 45 °C charge-discharge standard test protocols.

Figure 4.16 shows the appearance and dimension of NMC 3Ah cell under the test in the different kinds of temperatures in the climate chamber. Moreover, Figure 4.17 shows the test flow used for evaluating reliability in the Li-ion batteries based on the capacity fade indicator.

The capacity fades variation of the battery at 10 °C and 45 °C under standard test protocols are shown in Figure 4.18. According to Figure 4.18, the capacity at 45 °C and charge-discharge standard test protocols after 300 cycles are 2350 mAh and 85%, and 2251 mAh and 80%, respectively. The results describe the rate of changing capacity fade at different temperatures and under standard charge-discharge test protocols from the previous condition to the new condition during the battery's lifetime.

4.7 Conclusion

This chapter reviews the progress status of the active and passive TMS, aging, and reliability evaluation of the li-ion batteries. The liquid and air-based TMS is studied in-depth, and different solutions are presented to expand active cooling systems. The air cooling system benefits the simple design, low cost, and safe structure. Nevertheless, it suffers from low heat capacity and thermal efficiency. Liquid cooling TMS is a very operative cooling and promising method for stressful conditions. It benefits the high heat transfer coefficient of liquids. The thermal features of liquid cooling can be boosted using liquid metals, nanofluids, and boiling liquids due to their higher thermal conductivity. However, liquid cooling suffers high costs and leakage that need the proper sealing cover to avoid short circuits.

The PCM and heat pipe-based cooling systems are summarised, and advantages and challenges are discussed. New and recent techniques have been considered for passive and hybrid cooling systems. Using PCM and other active/passive cooling systems can increase thermal conductivity and decrease power consumption, which is the critical point for the development trend of PCM-based TMS. However, PCMs suffer leakage, chemical stability, and small phase change temperature area. Heat pipe-based TMS are efficient due to the high thermal conductivity, flexibility in shape, and compactness of the heat pipes. However, heat pipe cooling methods suffer the gravity, high cost, and complex design. The lifetime performance, battery degradation, and characterisation techniques of aging are considered as well.

References

1. F. H. Gandoman *et al.*, "Status and future perspectives of reliability assessment for electric vehicles," *Reliability Engineering & System Safety*, vol. 183, pp. 1–16, Mar. 2019, doi: 10.1016/J.RESS.2018.11.013.
2. Kalogiannis, Theodoros, *et al.* "Effects analysis on energy density optimisation and thermal efficiency enhancement of the air-cooled Li-ion battery modules." *Journal of Energy Storage* 48 (2022): 103847.
3. F. H. Gandoman *et al.*, "Chapter 16 - Reliability evaluation of Li-ion batteries for electric vehicles applications from the thermal perspectives," in *Uncertainties in Modern Power Systems*, A. F. Zobaa and S. H. E. Abdel Aleem, Eds. Academic Press, 2021, pp. 563–587. doi: https://doi.org/10.1016/B978-0-12-820491-7.00016-5.
4. H. Behi *et al.*, "Novel thermal management methods to improve the performance of the Li-ion batteries in high discharge current applications," *Energy*, p. 120165, 2021, doi: https://doi.org/10.1016/j.energy.2021.120165.
5. D. K. Sharma and A. Prabhakar, "A review on air cooled and air centric hybrid thermal management techniques for Li-ion battery packs in electric vehicles," *Journal of Energy Storage*, vol. 41, p. 102885, 2021, doi: https://doi.org/10.1016/j.est.2021.102885.
6. G. Zhao, X. Wang, M. Negnevitsky, and H. Zhang, "A review of air-cooling battery thermal management systems for electric and hybrid electric vehicles," *Journal of Power Sources*, vol. 501, p. 230001, 2021, doi: https://doi.org/10.1016/j.jpowsour.2021.230001.

7. G. Murali, G. S. N. Sravya, J. Jaya, and V. Naga Vamsi, "A review on hybrid thermal management of battery packs and it's cooling performance by enhanced PCM," *Renewable and Sustainable Energy Reviews*, vol. 150, p. 111513, 2021, doi: https://doi.org/10.1016/j.rser.2021.111513.
8. S. Chacko and S. Charmer, "Lithium-ion pack thermal modeling and evaluation of indirect liquid cooling for electric vehicle battery thermal management," in *Innovations in Fuel Economy and Sustainable Road Transport*, Woodhead Publishing, 2011, pp. 13–21. doi: https://doi.org/10.1533/9780857095879.1.13.
9. B. Sundén, "Chapter 6 - Thermal management of batteries," in *Hydrogen, Batteries and Fuel Cells*, B. Sundén, Ed. Academic Press, 2019, pp. 93–110. doi: https://doi.org/10.1016/B978-0-12-816950-6.00006-3.
10. R. Klinger and S. Jagsch, "Development of a detailed model for simulation and investigation of the thermal behaviour of electric vehicle traction batteries," in *Vehicle Thermal Management Systems Conference and Exhibition (VTMS10)*, Woodhead Publishing, 2011, pp. 299–307. doi: https://doi.org/10.1533/9780857095053.4.299.
11. A. K. Thakur *et al.*, "A state of art review and future viewpoint on advance cooling techniques for Lithium–ion battery system of electric vehicles," *Journal of Energy Storage*, vol. 32, p. 101771, 2020, doi: https://doi.org/10.1016/j.est.2020.101771.
12. M. Lu, X. Zhang, J. Ji, X. Xu, and Y. Zhang, "Research progress on power battery cooling technology for electric vehicles," *Journal of Energy Storage*, vol. 27, p. 101155, Feb. 2020, doi: 10.1016/J.EST.2019.101155.
13. Till Bunsen *et al.*, "Global EV Outlook 2019 to electric mobility," *OECD iea.org*, p. 232, 2019.
14. F. H. Gandoman, J. van Mierlo, A. Ahmadi, S. H. E. Abdel Aleem, and K. Chauhan, "Safety and reliability evaluation for electric vehicles in modern power system networks," *Distributed Energy Resources in Microgrids: Integration, Challenges and Optimisation*, pp. 389–404, Jan. 2019, doi: 10.1016/B978-0-12-817774-7.00015-6.
15. M. Akbarzadeh *et al.*, "A novel liquid cooling plate concept for thermal management of lithium-ion batteries in electric vehicles," *Energy Conversion and Management*, vol. 231, 113862, 2021, doi: 10.1016/j.enconman.2021.113862.
16. S. Khaleghi *et al.*, "Online health diagnosis of lithium-ion batteries based on nonlinear autoregressive neural network," *Applied Energy*, vol. 282, 116159, 2021, doi: 10.1016/j.apenergy.2020.116159.
17. T. Trigg, P. Telleen, R. Boyd, and F. Cuenot, "Global EV Outlook: Understanding the Electric Vehicle Landscape to 2020," *Iea*, no. April, pp. 1–41, 2013.
18. W. Lu, A. Jansen, D. Dees, P. Nelson, N. R. Veselka, and G. Henriksen, "High-energy electrode investigation for plug-in hybrid electric vehicles," *Journal of Power Sources*, vol. 196, no. 3, pp. 1537–1540, 2011, doi: https://doi.org/10.1016/j.jpowsour.2010.08.117.

19. R. Zhao, J. Liu, and J. Gu, "The effects of electrode thickness on the electrochemical and thermal characteristics of lithium ion battery," *Applied Energy*, vol. 139, pp. 220–229, 2015, doi: https://doi.org/10.1016/j.apenergy.2014.11.051.
20. D. Karimi, H. Behi, M. S. Hosen, J. Jaguemont, M. Berecibar, and J. Van Mierlo, "A compact and optimised liquid-cooled thermal management system for high power lithium-ion capacitors," *Applied Thermal Engineering*, vol. 185, p. 116449, 2021, doi: https://doi.org/10.1016/j.applthermaleng.2020.116449.
21. D. Karimi, H. Behi, J. Jaguemont, M. Berecibar, and J. Van Mierlo, "Optimised air-cooling thermal management system for high power lithium-ion capacitors," *energy*, vol. 1, no. 1, 2020.
22. D. Karimi, H. Behi, J. Van Mierlo, and M. Berecibar, "A Comprehensive Review of Lithium-Ion Capacitor Technology: Theory, Development, Modeling, Thermal Management Systems, and Applications," *Molecules*, vol. 27, no. 10, 3119, 2022, doi: 10.3390/molecules27103119.
23. F. Bahiraei, A. Fartaj, and G. A. Nazri, "Electrochemical-thermal Modeling to Evaluate Active Thermal Management of a Lithium-ion Battery Module," *Electrochimica Acta*, vol. 254, pp. 59–71, 2017, doi: 10.1016/j.electacta.2017.09.084.
24. H. Behi, T. Kalogiannis, M. Suresh Patil, J. Van Mierlo, and M. Berecibar, "A New Concept of Air Cooling and Heat Pipe for Electric Vehicles in Fast Discharging," *Energies*, vol. 14, no. 20, 2021, doi: 10.3390/en14206477.
25. S. Shahid and M. Agelin-Chaab, "Development and analysis of a technique to improve air-cooling and temperature uniformity in a battery pack for cylindrical batteries," *Thermal Science and Engineering Progress*, vol. 5, pp. 351–363, Mar. 2018, doi: 10.1016/J.TSEP.2018.01.003.
26. K. Chen, M. Song, W. Wei, and S. Wang, "Structure optimisation of parallel air-cooled battery thermal management system with U-type flow for cooling efficiency improvement," *Energy*, vol. 145, pp. 603–613, Feb. 2018, doi: 10.1016/j.energy.2017.12.110.
27. Z. Lu *et al.*, "Parametric study of forced air cooling strategy for lithium-ion battery pack with staggered arrangement," *Applied Thermal Engineering*, vol. 136, pp. 28–40, May 2018, doi: 10.1016/j.applthermaleng.2018.02.080.
28. K. Chen, M. Song, W. Wei, and S. Wang, "Design of the structure of battery pack in parallel air-cooled battery thermal management system for cooling efficiency improvement," *International Journal of Heat and Mass Transfer*, vol. 132, pp. 309–321, 2019, doi: 10.1016/j.ijheatmasstransfer.2018.12.024.
29. R. D. Jilte, R. Kumar, M. H. Ahmadi, and L. Chen, "Battery thermal management system employing phase change material with cell-to-cell air cooling," *Applied Thermal Engineering*, vol. 161, no. July, p. 114199, 2019, doi: 10.1016/j.applthermaleng.2019.114199.
30. X. Peng, C. Ma, A. Garg, N. Bao, and X. Liao, "Thermal performance investigation of an air-cooled lithium-ion battery pack considering the

inconsistency of battery cells," *Applied Thermal Engineering*, vol. 153, pp. 596–603, May 2019, doi: 10.1016/J.APPLTHERMALENG.2019.03.042.
31. X. Li, F. He, G. Zhang, Q. Huang, and D. Zhou, "Experiment and simulation for pouch battery with silica cooling plates and copper mesh based air cooling thermal management system," *Applied Thermal Engineering*, vol. 146, pp. 866–880, Jan. 2019, doi: 10.1016/J.APPLTHERMALENG.2018.10.061.
32. E. Jiaqiang *et al.*, "Effects of the different air cooling strategies on cooling performance of a lithium-ion battery module with baffle," *Applied Thermal Engineering*, vol. 144, no. May, pp. 231–241, 2018, doi: 10.1016/j.applthermaleng.2018.08.064.
33. K. Chen, W. Wu, F. Yuan, L. Chen, and S. Wang, "Cooling efficiency improvement of air-cooled battery thermal management system through designing the flow pattern," *Energy*, vol. 167, pp. 781–790, Jan. 2019, doi: 10.1016/J.ENERGY.2018.11.011.
34. A. A. H. Akinlabi and D. Solyali, "Configuration, design, and optimisation of air-cooled battery thermal management system for electric vehicles: A review," *Renewable and Sustainable Energy Reviews*, vol. 125, p. 109815, 2020, doi: https://doi.org/10.1016/j.rser.2020.109815.
35. X. Yu, Z. Lu, L. Zhang, L. Wei, X. Cui, and L. Jin, "Experimental study on transient thermal characteristics of stagger-arranged lithium-ion battery pack with air cooling strategy," *International Journal of Heat and Mass Transfer*, vol. 143, p. 118576, Nov. 2019, doi: 10.1016/j.ijheatmasstransfer.2019.118576.
36. Y. Fan, Y. Bao, C. Ling, Y. Chu, X. Tan, and S. Yang, "Experimental study on the thermal management performance of air cooling for high energy density cylindrical lithium-ion batteries," *Applied Thermal Engineering*, vol. 155, pp. 96–109, Jun. 2019, doi: 10.1016/J.APPLTHERMALENG.2019.03.157.
37. F. Zhang, A. Lin, P. Wang, and P. Liu, "Optimisation design of a parallel air-cooled battery thermal management system with spoilers," *Applied Thermal Engineering*, vol. 182, p. 116062, Jan. 2021, doi: 10.1016/j.applthermaleng.2020.116062.
38. K. Chen, Y. Chen, Y. She, M. Song, S. Wang, and L. Chen, "Construction of effective symmetrical air-cooled system for battery thermal management," *Applied Thermal Engineering*, vol. 166, p. 114679, Feb. 2020, doi: 10.1016/J.APPLTHERMALENG.2019.114679.
39. L. Cheng, A. Garg, A. K. Jishnu, and L. Gao, "Surrogate based multi-objective design optimisation of lithium-ion battery air-cooled system in electric vehicles," *Journal of Energy Storage*, vol. 31, p. 101645, 2020, doi: https://doi.org/10.1016/j.est.2020.101645.
40. W. Yang, F. Zhou, H. Zhou, and Y. Liu, "Thermal performance of axial air cooling system with bionic surface structure for cylindrical lithium-ion battery module," *International Journal of Heat and Mass Transfer*, vol. 161, p. 120307, 2020, doi: https://doi.org/10.1016/j.ijheatmasstransfer.2020.120307.
41. K. Chen, S. Wang, M. Song, and L. Chen, "Configuration optimisation of battery pack in parallel air-cooled battery thermal management system using an

optimisation strategy," *Applied Thermal Engineering*, vol. 123, pp. 177–186, Aug. 2017, doi: 10.1016/j.applthermaleng.2017.05.060.
42. D. C. Erb, S. Kumar, E. Carlson, I. M. Ehrenberg, and S. E. Sarma, "Analytical methods for determining the effects of lithium-ion cell size in aligned air-cooled battery packs," *Journal of Energy Storage*, vol. 10, pp. 39–47, 2017, doi: 10.1016/j.est.2016.12.003.
43. H. Behi *et al.*, "Aluminum Heat Sink Assisted Air-Cooling Thermal Management System for High Current Applications in Electric Vehicles," in *2020 AEIT International Conference of Electrical and Electronic Technologies for Automotive (AEIT AUTOMOTIVE)*, Nov. 2020, pp. 1–6.
44. D. Karimi, H. Behi, J. Jaguemont, M. El Baghdadi, J. Van Mierlo, and O. Hegazy, "Thermal Concept Design of MOSFET Power Modules in Inverter Subsystems for Electric Vehicles," 2019. doi: 10.1109/ICPES47639.2019.9105437.
45. M. Akbarzadeh *et al.*, "A comparative study between air cooling and liquid cooling thermal management systems for a high-energy lithium-ion battery module," *Applied Thermal Engineering*, vol. 198, p. 117503, 2021, doi: https://doi.org/10.1016/j.applthermaleng.2021.117503.
46. M. Akbarzadeh *et al.*, "Experimental and numerical thermal analysis of a lithium-ion battery module based on a novel liquid cooling plate embedded with phase change material," *Journal of Energy Storage*, vol. 50, 104673, 2022, doi: 10.1016/j.est.2022.104673.
47. Z. Y. Jiang and Z. G. Qu, "Lithium–ion battery thermal management using heat pipe and phase change material during discharge–charge cycle: A comprehensive numerical study," *Applied Energy*, vol. 242, no. February, pp. 378–392, 2019, doi: 10.1016/j.apenergy.2019.03.043.
48. DOBER, "COOLING ELECTRIC VEHICLES," *https://www.dober.com/electric-vehicle-cooling-systems#electric_vehicle_thermal_management_system.*
49. H. Zhou, F. Zhou, Q. Zhang, Q. Wang, and Z. Song, "Thermal management of cylindrical lithium-ion battery based on a liquid cooling method with half-helical duct," *Applied Thermal Engineering*, vol. 162, p. 114257, Nov. 2019, doi: 10.1016/j.applthermaleng.2019.114257.
50. Y. Wang, G. Zhang, and X. Yang, "Optimisation of liquid cooling technology for cylindrical power battery module," *Applied Thermal Engineering*, vol. 162, p. 114200, Nov. 2019, doi: 10.1016/J.APPLTHERMALENG.2019.114200.
51. M. Al-Zareer, I. Dincer, and M. A. Rosen, "A novel approach for performance improvement of liquid to vapor based battery cooling systems," *Energy Conversion and Management*, vol. 187, pp. 191–204, May 2019, doi: 10.1016/j.enconman.2019.02.063.
52. J. E *et al.*, "Orthogonal experimental design of liquid-cooling structure on the cooling effect of a liquid-cooled battery thermal management system," *Applied Thermal Engineering*, vol. 132, pp. 508–520, Mar. 2018, doi: 10.1016/J.APPLTHERMALENG.2017.12.115.
53. S. Chen, X. Peng, N. Bao, and A. Garg, "A comprehensive analysis and optimisation process for an integrated liquid cooling plate for a prismatic

lithium-ion battery module," *Applied Thermal Engineering*, vol. 156, pp. 324–339, 2019, doi: https://doi.org/10.1016/j.applthermaleng.2019.04.089.

54. M. Yates, M. Akrami, and A. A. Javadi, "Analysing the performance of liquid cooling designs in cylindrical lithium-ion batteries," *Journal of Energy Storage*, p. 100913, 2019, doi: https://doi.org/10.1016/j.est.2019.100913.
55. M. Al-Zareer, I. Dincer, and M. A. Rosen, "Comparative assessment of new liquid-to-vapor type battery cooling systems," *Energy*, vol. 188, p. 116010, Dec. 2019, doi: 10.1016/j.energy.2019.116010.
56. J. Cao, Z. Ling, X. Fang, and Z. Zhang, "Delayed liquid cooling strategy with phase change material to achieve high temperature uniformity of Li-ion battery under high-rate discharge," *Journal of Power Sources*, vol. 450, p. 227673, Feb. 2020, doi: 10.1016/J.JPOWSOUR.2019.227673.
57. Y. Lai, W. Wu, K. Chen, S. Wang, and C. Xin, "A compact and lightweight liquid-cooled thermal management solution for cylindrical lithium-ion power battery pack," *International Journal of Heat and Mass Transfer*, vol. 144, p. 118581, Dec. 2019, doi: 10.1016/J.IJHEATMASSTRANSFER.2019.118581.
58. Z. Zhu, X. Wu, H. Zhang, Y. Guo, and G. Wu, "Multi-objective optimisation of a liquid cooled battery module with collaborative heat dissipation in both axial and radial directions," *International Journal of Heat and Mass Transfer*, vol. 155, p. 119701, 2020, doi: https://doi.org/10.1016/j.ijheatmasstransfer.2020.119701.
59. W. Yang, F. Zhou, H. Zhou, Q. Wang, and J. Kong, "Thermal performance of cylindrical Lithium-ion battery thermal management system integrated with mini-channel liquid cooling and air cooling," *Applied Thermal Engineering*, p. 115331, Apr. 2020, doi: 10.1016/J.APPLTHERMALENG.2020.115331.
60. Y. Huang, S. Wang, Y. Lu, R. Huang, and X. Yu, "Study on a liquid cooled battery thermal management system pertaining to the transient regime," *Applied Thermal Engineering*, vol. 180, p. 115793, Nov. 2020, doi: 10.1016/j.applthermaleng.2020.115793.
61. Z. Tang, S. Wang, Z. Liu, and J. Cheng, "Numerical analysis of temperature uniformity of a liquid cooling battery module composed of heat-conducting blocks with gradient contact surface angles," *Applied Thermal Engineering*, vol. 178, p. 115509, 2020, doi: https://doi.org/10.1016/j.applthermaleng.2020.115509.
62. X. Wu, Z. Zhu, H. Zhang, S. Xu, Y. Fang, and Z. Yan, "Structural optimisation of lightweight battery module based on hybrid liquid cooling with high latent heat PCM," *International Journal of Heat and Mass Transfer*, vol. 163, p. 120495, Dec. 2020, doi: 10.1016/j.ijheatmasstransfer.2020.120495.
63. X. Yuan, A. Tang, C. Shan, Z. Liu, and J. Li, "Experimental investigation on thermal performance of a battery liquid cooling structure coupled with heat pipe," *Journal of Energy Storage*, vol. 32, p. 101984, Dec. 2020, doi: 10.1016/j.est.2020.101984.
64. H. Wang, T. Tao, J. Xu, X. Mei, X. Liu, and P. Gou, "Cooling capacity of a novel modular liquid-cooled battery thermal management system for cylindrical

lithium ion batteries," *Applied Thermal Engineering*, vol. 178, p. 115591, 2020, doi: https://doi.org/10.1016/j.applthermaleng.2020.115591.

65. Y. Xu, X. Li, X. Liu, Y. Wang, X. Wu, and D. Zhou, "Experiment investigation on a novel composite silica gel plate coupled with liquid-cooling system for square battery thermal management," *Applied Thermal Engineering*, p. 116217, Oct. 2020, doi: 10.1016/j.applthermaleng.2020.116217.
66. H. Behi *et al.*, "Enhancement of the Thermal Energy Storage Using Heat-Pipe-Assisted Phase Change Material," *Energies*, vol. 14, no. 19, 2021, doi: 10.3390/en14196176.
67. D. Karimi, H. Behi, M. Akbarzadeh, J. Van Mierlo, and M. Berecibar, "Holistic 1D Electro-Thermal Model Coupled to 3D Thermal Model for Hybrid Passive Cooling System Analysis in Electric Vehicles," *Energies*, vol. 14, no. 18, 2021, doi: 10.3390/en14185924.
68. D. Karimi, H. Behi, J. Van Mierlo, and M. Berecibar, "Novel Hybrid Thermal Management System for High-Power Lithium-Ion Module for Electric Vehicles: Fast Charging Applications," *World Electric Vehicle Journal*, vol. 13, no. 5, 86, 2022, doi: 10.3390/wevj13050086.
69. H. Behi, "Experimental and numerical study on heat pipe assisted PCM storage system." 2015.
70. D. Karimi *et al.*, "Thermal performance enhancement of phase change material using aluminum-mesh grid foil for lithium-capacitor modules," *Journal of Energy Storage*, vol. 30, p. 101508, 2020, doi: https://doi.org/10.1016/j.est.2020.101508.
71. H. Behi, M. Ghanbarpour, and M. Behi, "Investigation of PCM-assisted heat pipe for electronic cooling," *Applied Thermal Engineering*, vol. 127, pp. 1132–1142, Dec. 2017, doi: 10.1016/J.APPLTHERMALENG.2017.08.109.
72. S. Mirmohammadi and M. Behi, "Investigation on Thermal Conductivity, Viscosity and Stability of Nanofluids," p. 140, 2012.
73. S. A. Mirmohammadi, M. Behi, Y. Gan, and L. Shen, "Particle-shape-, temperature-, and concentration-dependent thermal conductivity and viscosity of nanofluids," *Phys. Rev. E*, vol. 99, no. 4, p. 43109, Apr. 2019, doi: 10.1103/PhysRevE.99.043109.
74. M. Behi, S. A. Mirmohammadi, A. B. Suma, and B. E. Palm, "Optimized Energy Recovery in Line With Balancing of an ATES." Jul. 28, 2014. doi: 10.1115/POWER2014-32017.
75. M. Behi *et al.*, "Experimental and numerical investigation on hydrothermal performance of nanofluids in micro-tubes," *Energy*, vol. 193, p. 116658, Feb. 2020, doi: 10.1016/J.ENERGY.2019.116658.
76. S. Al Hallaj and J. R. Selman, "Novel thermal management system for electric vehicle batteries using phase-change material," *Journal of the Electrochemical Society*, vol. 147, no. 9, pp. 3231–3236, 2000, doi: 10.1149/1.1393888.
77. W. Wu *et al.*, "An innovative battery thermal management with thermally induced flexible phase change material," *Energy Conversion and Management*, vol. 221, no. March, p. 113145, 2020, doi: 10.1016/j.enconman.2020.113145.

78. M. Behi, S. A. Mirmohammadi, M. Ghanbarpour, H. Behi, and B. Palm, "Evaluation of a novel solar driven sorption cooling/heating system integrated with PCM storage compartment," *Energy*, vol. 164, pp. 449–464, Dec. 2018, doi: 10.1016/J.ENERGY.2018.08.166.
79. H. Behi *et al.*, "PCM assisted heat pipe cooling system for the thermal management of an LTO cell for high-current profiles," *Case Studies in Thermal Engineering*, vol. 25, p. 100920, 2021, doi: https://doi.org/10.1016/j.csite.2021.100920.
80. H. Yang, H. Zhang, Y. Sui, and C. Yang, "Numerical analysis and experimental visualisation of phase change material melting process for thermal management of cylindrical power battery," *Applied Thermal Engineering*, vol. 128, pp. 489–499, Jan. 2018, doi: 10.1016/J.APPLTHERMALENG.2017.09.022.
81. S. A. Khateeb, M. M. Farid, J. R. Selman, and S. Al-Hallaj, "Design and simulation of a lithium-ion battery with a phase change material thermal management system for an electric scooter," *Journal of Power Sources*, vol. 128, no. 2, pp. 292–307, Apr. 2004, doi: 10.1016/J.JPOWSOUR.2003.09.070.
82. S. Nemati and G. Pircheraghi, "Fabrication of a form-stable phase change material with green fatty acid and recycled silica nanoparticles from spent lead-acid battery separators with enhanced thermal conductivity," *Thermochimica Acta*, vol. 693, p. 178781, Nov. 2020, doi: 10.1016/j.tca.2020.178781.
83. Q. Huang, J. Deng, X. Li, G. Zhang, and F. Xu, "Experimental investigation on thermally induced aluminum nitride based flexible composite phase change material for battery thermal management," *Journal of Energy Storage*, vol. 32, p. 101755, Dec. 2020, doi: 10.1016/j.est.2020.101755.
84. H. Li, X. Xiao, Y. Wang, C. Lian, Q. Li, and Z. Wang, "Performance investigation of a battery thermal management system with microencapsulated phase change material suspension," *Applied Thermal Engineering*, vol. 180, p. 115795, Nov. 2020, doi: 10.1016/j.applthermaleng.2020.115795.
85. J. Zhang *et al.*, "Experimental investigation of the flame retardant and form-stable composite phase change materials for a power battery thermal management system," *Journal of Power Sources*, vol. 480, p. 229116, Dec. 2020, doi: 10.1016/j.jpowsour.2020.229116.
86. Y. Lv, G. Liu, G. Zhang, and X. Yang, "A novel thermal management structure using serpentine phase change material coupled with forced air convection for cylindrical battery modules," *Journal of Power Sources*, vol. 468, p. 228398, Aug. 2020, doi: 10.1016/j.jpowsour.2020.228398.
87. J. Li and H. Zhang, "Thermal characteristics of power battery module with composite phase change material and external liquid cooling," *International Journal of Heat and Mass Transfer*, vol. 156, p. 119820, Aug. 2020, doi: 10.1016/j.ijheatmasstransfer.2020.119820.
88. S. Lei, Y. Shi, and G. Chen, "A lithium-ion battery-thermal-management design based on phase-change-material thermal storage and spray cooling," *Applied Thermal Engineering*, vol. 168, p. 114792, Mar. 2020, doi: 10.1016/J.APPLTHERMALENG.2019.114792.

89. J. Cao *et al.*, "Mini-channel cold plate with nano phase change material emulsion for Li-ion battery under high-rate discharge," *Applied Energy*, vol. 279, p. 115808, Dec. 2020, doi: 10.1016/j.apenergy.2020.115808.
90. M. Kiani, S. Omiddezyani, E. Houshfar, S. R. Miremadi, M. Ashjaee, and A. Mahdavi Nejad, "Lithium-ion battery thermal management system with Al2O3/AgO/CuO nanofluids and phase change material," *Applied Thermal Engineering*, vol. 180, p. 115840, Nov. 2020, doi: 10.1016/j.applthermaleng.2020.115840.
91. G. R. Molaeimanesh, S. M. Mirfallah Nasiry, and M. Dahmardeh, "Impact of configuration on the performance of a hybrid thermal management system including phase change material and water-cooling channels for Li-ion batteries," *Applied Thermal Engineering*, vol. 181, p. 116028, Nov. 2020, doi: 10.1016/j.applthermaleng.2020.116028.
92. V. G. Choudhari, A. S. Dhoble, and S. Panchal, "Numerical analysis of different fin structures in phase change material module for battery thermal management system and its optimisation," *International Journal of Heat and Mass Transfer*, vol. 163, p. 120434, Dec. 2020, doi: 10.1016/j.ijheatmasstransfer.2020.120434.
93. Z. Wencan, L. Zhicheng, Y. Xiuxing, and L. Guozhi, "Avoiding thermal runaway propagation of lithium-ion battery modules by using hybrid phase change material and liquid cooling," *Applied Thermal Engineering*, p. 116380, Nov. 2020, doi: 10.1016/j.applthermaleng.2020.116380.
94. Z. Liu, J. Huang, M. Cao, G. Jiang, Q. Yan, and J. Hu, "Experimental Study on the Thermal Management of Batteries Based on the Coupling of Composite Phase Change Materials and Liquid Cooling," *Applied Thermal Engineering*, p. 116415, Dec. 2020, doi: 10.1016/j.applthermaleng.2020.116415.
95. M. A. Bamdezh, G. R. Molaeimanesh, and S. Zanganeh, "Role of foam anisotropy used in the phase-change composite material for the hybrid thermal management system of lithium-ion battery," *Journal of Energy Storage*, vol. 32, p. 101778, Dec. 2020, doi: 10.1016/j.est.2020.101778.
96. J. Zhang *et al.*, "Characterisation and experimental investigation of aluminum nitride-based composite phase change materials for battery thermal management," *Energy Conversion and Management*, vol. 204, p. 112319, Jan. 2020, doi: 10.1016/j.enconman.2019.112319.
97. N. Joshy, M. Hajiyan, A. R. M. Siddique, S. Tasnim, H. Simha, and S. Mahmud, "Experimental investigation of the effect of vibration on phase change material (PCM) based battery thermal management system," *Journal of Power Sources*, vol. 450, p. 227717, Feb. 2020, doi: 10.1016/j.jpowsour.2020.227717.
98. F. Wang, J. Cao, Z. Ling, Z. Zhang, and X. Fang, "Experimental and simulative investigations on a phase change material nano-emulsion-based liquid cooling thermal management system for a lithium-ion battery pack," *Energy*, vol. 207, p. 118215, Sep. 2020, doi: 10.1016/j.energy.2020.118215.
99. H. Behi, D. Karimi, J. Jaguemont, M. Berecibar, and J. Van Mierlo, "Experimental study on cooling performance of flat heat pipe for lithium-ion

battery at various inclination angels," *Energy Perspectives*, vol. 1, no. 1, pp. 77–92, 2020.
100. D. Karimi *et al.*, "A hybrid thermal management system for high power lithium-ion capacitors combining heat pipe with phase change materials," *Heliyon*, vol. 7, no. 8, p. e07773, 2021, doi: https://doi.org/10.1016/j.heliyon.2021.e07773.
101. Y. Li, H. Guo, F. Qi, Z. Guo, M. Li, and L. Bertling Tjernberg, "Investigation on liquid cold plate thermal management system with heat pipes for LiFePO4 battery pack in electric vehicles," *Applied Thermal Engineering*, vol. 185, p. 116382, 2021, doi: https://doi.org/10.1016/j.applthermaleng.2020.116382.
102. H. Behi *et al.*, "Thermal management analysis using heat pipe in the high current discharging of lithium-ion battery in electric vehicles," *Journal of Energy Storage*, vol. 32, p. 101893, 2020, doi: https://doi.org/10.1016/j.est.2020.101893.
103. J. Liang, Y. Gan, Y. Li, M. Tan, and J. Wang, "Thermal and electrochemical performance of a serially connected battery module using a heat pipe-based thermal management system under different coolant temperatures," *Energy*, vol. 189, p. 116233, Dec. 2019, doi: 10.1016/J.ENERGY.2019.116233.
104. Y. Gan, J. Wang, J. Liang, Z. Huang, and M. Hu, "Development of thermal equivalent circuit model of heat pipe-based thermal management system for a battery module with cylindrical cells," *Applied Thermal Engineering*, vol. 164, p. 114523, Jan. 2020, doi: 10.1016/J.APPLTHERMALENG.2019.114523.
105. H. Behi *et al.*, "A new concept of thermal management system in Li-ion battery using air cooling and heat pipe for electric vehicles," *Applied Thermal Engineering*, p. 115280, Apr. 2020, doi: 10.1016/J.APPLTHERMALENG.2020.115280.
106. H. Behi *et al.*, "Heat pipe air-cooled thermal management system for lithium-ion batteries: High power applications," *Applied Thermal Engineering*, p. 116240, 2020, doi: https://doi.org/10.1016/j.applthermaleng.2020.116240.
107. N. Putra, A. F. Sandi, B. Ariantara, N. Abdullah, and T. M. Indra Mahlia, "Performance of beeswax phase change material (PCM) and heat pipe as passive battery cooling system for electric vehicles," *Case Studies in Thermal Engineering*, vol. 21, p. 100655, 2020, doi: https://doi.org/10.1016/j.csite.2020.100655.
108. W. Zhang, J. Qiu, X. Yin, and D. Wang, "A novel heat pipe assisted separation type battery thermal management system based on phase change material," *Applied Thermal Engineering*, vol. 165, p. 114571, 2020, doi: https://doi.org/10.1016/j.applthermaleng.2019.114571.
109. Q. Huang, X. Li, G. Zhang, J. Zhang, F. He, and Y. Li, "Experimental investigation of the thermal performance of heat pipe assisted phase change material for battery thermal management system," *Applied Thermal Engineering*, vol. 141, no. June, pp. 1092–1100, 2018, doi: 10.1016/j.applthermaleng.2018.06.048.
110. D. Karimi *et al.*, "An Experimental Study on Thermal Performance of Graphite-Based Phase-Change Materials for High-Power Batteries," *Energies*, vol. 15, no. 7, 2515, 2022, doi: 10.3390/en15072515.

111. S. Khaleghi *et al.*, "Developing an online data-driven approach for prognostics and health management of lithium-ion batteries," *Applied Energy*, vol. 308, 118348, 2022, doi: 10.1016/j.apenergy.2021.118348.
112. M. S. Hosen *et al.*, "Electro-aging model development of nickel-manganese-cobalt lithium-ion technology validated with light and heavy-duty real-life profiles," *Journal of Energy Storage*, vol. 28, 101265, 2020, doi: 10.1016/j.est.2020.101265.
113. D. Karimi *et al.*, "Modular Methodology for Developing Comprehensive Active and Passive Thermal Management Systems for Electric Vehicle," *Ph.D. thesis, 2022, Vrije Universiteit Brussel, Belgium.*
114. H. Behi *et al.*, "Advanced hybrid thermal management system for LTO battery module under fast charging," *Case Studies in Thermal Engineering*, vol. 33101938, 2022, doi: 10.1016/j.csite.2022.101938.
115. D. Karimi *et al.*, "Lithium-ion capacitor lifetime extension through an optimal thermal management system for smart grid applications," *Energies*, vol. 14, no. 10, 2907, 2021, doi: 10.3390/en14102907.
116. H. Behi *et al.*, "Comprehensive passive thermal management systems for electric vehicles," *Energies*, vol. 14, no. 13, 3881, 2021, doi: 10.3390/en14133881.
117. J. Jaguemont *et al.*, "Investigation of a passive thermal management system for lithium-ion capacitors," *IEEE Transactions on Vehicular Technology,* vol. 68, no. 11, pp. 10518-10524, 2019, doi: 10.1109/TVT.2019.2939632.
118. D. Karimi *et al.*, "Investigation of extruded heat sink assisted air cooling system for lithium-ion capacitor batteries," *In Proceedings of the International Conference on Renewable Energy Systems and Environmental Engineering, Brussels, Belgium, 18 July–20 September 2020; Global Publisher: Brussels, Belgium, 2020; pp. 1–6.*
119. D. Karimi *et al.*, "Passive cooling based battery thermal management using phase change materials for electric vehicles," *In Proceedings of the EVS33 International Electric Vehicle Symposium, Portland, OR, USA, 14–17 June 2020; pp. 1–12.*
120. S. Moeller *et al.*, "Application Considerations for Double Sided Cooled Modules in Automotive Environment," *In Proceedings of the CIPS 2020; 11th International Conference on Integrated Power Electronics Systems, Berlin, Germany, 24–26 March 2020; VDE: Berlin, Germany, 2020.*
121. MS. Hosen *et al.*, "Twin-model framework development for a comprehensive battery lifetime prediction validated with a realistic driving profile," *Energy Science & Engineering*, Vol 9, no. 11, pp. 2191-2201, 2021, doi: 10.1002/ese3.973.
122. D. Karimi *et al.*, "Optimisation of 1D/3D Electro-Thermal Model for Liquid-Cooled Lithium-Ion Capacitor Module in High Power Applications," Electricity, Vol 2, no. 4, pp. 503-523, 2021, doi: 10.3390/electricity2040030.
123. D. Karimi *et al.*, "A Novel Air-Cooled Thermal Management Approach towards High-Power Lithium-Ion Capacitor Module for Electric Vehicles," *Energies*, Vol 14, no. 21, 7150, 2021, doi: 10.3390/en14217150.

124. D. Y. Jung, B. H. Lee, and S. W. Kim, "Development of battery management system for nickel-metal hydride batteries in electric vehicle applications," *Journal of Power Sources*, vol. 109, no. 1, pp. 1–10, 2002, doi: 10.1016/S0378-7753(02)00020-4.
125. J. Chatzakis, K. Kalaitzakis, N. C. Voulgaris, and S. N. Manias, "Designing a new generalised battery management system," *IEEE Transactions on Industrial Electronics*, vol. 50, no. 5, pp. 990–999, 2003, doi: 10.1109/TIE.2003.817706.
126. I. Sefik, D. A. Asfani, and T. Hiyama, "Simulation-based analysis of short circuit fault in parallel-series type hybrid electric vehicle," *2011 International Conference on Advanced Power System Automation and Protection*, pp. 2045–2049, 2011, doi: 10.1109/APAP.2011.6180687.
127. X. Xu and N. Chen, "A state-space-based prognostics model for lithium-ion battery degradation," *Reliability Engineering and System Safety*, vol. 159, no. October 2016, pp. 47–57, 2017, doi: 10.1016/j.ress.2016.10.026.
128. B. Akin and S. Choi, "Machines in Hybrid Electric," *IEEE Signal Processing Magazine*, no. May, 2012.
129. W. Zhao, M. Cheng, K. T. Chau, R. Cao, and J. Ji, "Remedial injected-harmonic-current operation of redundant flux-switching permanent-magnet motor drives," *IEEE Transactions on Industrial Electronics*, vol. 60, no. 1, pp. 181–189, 2013, doi: 10.1109/TIE.2012.2186107.
130. A. Affanni, A. Bellini, G. Franceschini, P. Guglielmi, and C. Tassoni, "Battery choice and management for new-generation electric vehicles," *IEEE Transactions on Industrial Electronics*, vol. 52, no. 5, pp. 1343–1349, 2005, doi: 10.1109/TIE.2005.855664.
131. K. T. Chau and Z. Wang, "Overview of power electronic drives for electric vehicles," *Technology*, vol. 2, pp. 737–761, 2005.
132. X. Zhu, K. T. Chau, M. Cheng, and C. Yu, "Design and control of a flux-controllable stator-permanent magnet brushless motor drive," *Journal of Applied Physics*, vol. 103, no. 7, 2008, doi: 10.1063/1.2838325.
133. B. Lu and S. K. Sharma, "A literature review of IGBT fault diagnostic and protection methods for power inverters," *IEEE Transactions on Industry Applications*, vol. 45, no. 5, pp. 1770–1777, 2009, doi: 10.1109/TIA.2009.2027535.
134. "https://batterybro.com/blogs/18650-wholesale-battery-reviews/57179459-lg-hg2-review-20a-3000mah."

5

Fault Detection and Isolation in Electric Vehicle Powertrain

Gbanaibolou Jombo[1,2*] and Yu Zhang[3]

[1]*Centre for Engineering Research, School of Physics, Engineering and Computer Science, University of Hertfordshire, Hatfield, UK*
[2]*Centre for Climate Change Research (C3R), University of Hertfordshire, Hatfield, UK*
[3]*Department of Aeronautical and Automotive Engineering, Loughborough University, Loughborough, UK*

Abstract

The powertrain of an electric vehicle (EV) consists mainly of the battery, electric motor and power electronics. The safe and reliable operation of the electric vehicle depends on their fault-free operation. Fault detection and isolation methods work on the premise that small changes as a result of faults affecting a system causes variation in its operational response. This property can be used for the detection of such faults and their severity. This chapter discusses methods for detection and isolation of faults in electric vehicle powertrain components. Powertrain configuration and technologies are identified. Battery technology such as Lithium-ion batteries have gained a significant application as energy storage source in electric vehicles due to their high energy and power density, long lifespan, and low self-discharge performance under extreme temperatures. Model-based approaches are discussed for the determination of battery state of charge, state of health and effect of accelerated degradation. Fault detection in electric motor is considered. Brushless asynchronous induction motor, brushed externally excited synchronous motor and brushless permanent magnet synchronous motor are the options adopted for the electric vehicle powertrain. A signal processing-based approach such as the motor current signature analysis is explored for detection of broken rotor bar, shorten stator windings, air gap eccentricity, bearing failure and load variation effects. Lastly, fault detection in power electronics is explored.

**Corresponding author*: g.jombo@herts.ac.uk

Pedram Asef, Sanjeevikumar Padmanaban, and Andrew Lapthorn (eds.) Modern Automotive Electrical Systems, (193–226) © 2023 Scrivener Publishing LLC

Electric vehicle electric components need complex electronics to control them. These come in the form of a power electronics module (PEM), and an inverter, which can be integral with the PEM or the electric motor itself. Inverters provide the interface between an alternating current electric component and the direct current battery. The current focus for electric vehicle power electronics is controllable solid-state switches such as insulated gate bipolar transistor. For these power drives, the major faults are open switch fault and short switch fault. Signal processing-based approaches are considered for detection of these faults.

Keywords: Electric vehicle fault diagnosis, battery fault diagnosis, electric motor fault diagnosis, power electronics fault diagnosis

5.1 Introduction

An electric vehicle (EV) is a vehicle that is powered, at least in part, by an electric energy source. Based on the type and combination of energy sources used in an EV, the following configurations are widely adopted: battery electric vehicle (BEV), fuel cell electric vehicle (FCEV) and hybrid-electric vehicle (HEV) [1, 2]. The story of the EV is one of rebirth. EVs first came into prominence during the period 1897-1900, where they secured 28% of the automotive vehicle market; however, they met their demise in favour of the internal combustion engine (ICE) vehicle due to low oil prices [1, 3]. In recent times, the tide has turned in favour of the EV, due mainly to government legislation on the impact of climate change from greenhouse gas emission from fossil fuel-based ICE vehicles. As governments and industry alike craft a roadmap for decarbonising the transport sector by 2030, the EV would play an important role in achieving this. As such, the safe and reliable operation of electric vehicles becomes relevant. Fault detection and isolation (FDI) techniques provides the necessary approaches to ensure fault tolerant operation of electric vehicle powertrain components. The focus of this chapter is to review FDI techniques applied to EV powertrain components, highlighting the fundamental theories and applications.

5.1.1 EV Powertrain Configurations

An electric vehicle powertrain can take on different configurations based on the type and combination of energy sources adopted. Basically, EV powertrain components consist of an energy source (e.g., battery, fuel cell, ICE, ultra-capacitor, flywheel), and drivetrain components that ensures the conversion of electrical energy to motive force at the wheels (i.e., electric motor and power electronics). The various combinations of these powertrain components, especially as it relates to the energy sources, leads to three broad types of EV: battery electric vehicle, hybrid electric vehicle and fuel cell electric vehicle. Table 5.1 compares the various EV powertrain configurations.

Table 5.1 Comparison of EV powertrain configurations [2].

Types of EV	Battery EV	Hybrid EV	Fuel cell EV
Propulsion	• Electric motor drives	• Electric motor drives • Internal combustion engines	• Electric motor drives
Energy system	• Battery • Ultracapacitor	• Battery • Ultracapacitor • ICE generating unit	• Fuel cells
Energy source and infrastructure	• Electric grid charging facilities	• Gasoline stations • Electric grid charging facilities (optional)	• Hydrogen • Methanol or gasoline • Ethanol
Characteristics	• Zero emission • Independence on crude oils • 100-200 km short range • High initial cost • Commercially available	• Very low emission • Long driving range • Dependence on crude oils • Complex • Commercially available	• Zero emission or ultra-low emission • High energy efficiency • Independence on crude oils • Satisfied driving range • High cost now • Under development
Major issue	• Battery and battery management • High performance propulsion • Charging facilities	• Managing multiple energy sources • Dependent on driving cycle • Battery sizing and management	• Fuel cell cost • Fuel processor • Fuelling system

5.1.1.1 *Battery Electric Vehicle (BEV)*

BEV runs on the battery as the only energy source. Figure 5.1 shows typical configurations of BEV powertrain which can be adopted based on performance, compactness, weight, and cost requirements.

a) This configuration lends itself from a converted conventional ICE powertrain for a longitudinal front engine, front wheel drive. It consists of an electric motor, clutch, gearbox and differential. This setup is heavy, although its advantage is the ability to select different gear ratio due to the presence of the clutch and gearbox. As such, high torque and low speed requirement can be met at a low gear ratio, and low torque and highspeed requirement can be met at a higher gear ratio.

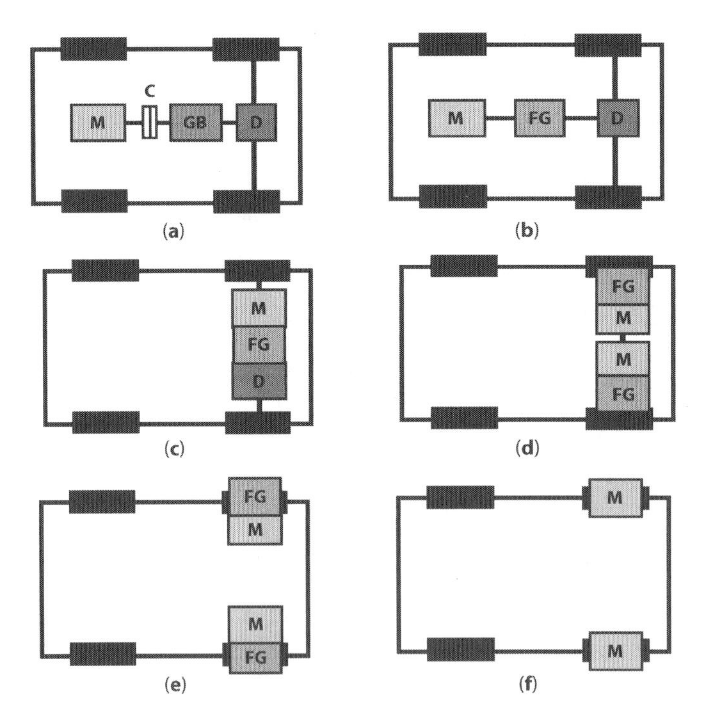

Figure 5.1 Typical BEV powertrain configurations: (a) BEV powertrain based on converted conventional ICE powertrain, (b) BEV powertrain with fixed gearing to replace clutch and gearbox, (c) BEV powertrain with integrated power assembly unit using mechanical differential, (d) BEV powertrain with integrated power assembly unit using electronic differential, (e) BEV powertrain for in-wheel configuration using motor and fixed gearing and (f) BEV powertrain for in-wheel configuration using only motor [2].

b) This configuration reduces the weight of the BEV powertrain by replacing the clutch and gearbox with a fixed gearing. It still preserves the mechanical differential as a means to enable both drive wheels to operate at different speeds.
c) This configuration is widely adopted by BEV. It takes the format of an integrated power assembly unit consisting of electric motor, fixed gearing and differential. However, unlike Figure 5.1(b), its arrangement takes the form similar to transverse front engine, front wheel drive.
d) This configuration eliminates the mechanical differential component with an electronic approach. It consists of dual electric motor with fixed gearing driving each wheel. The speed of each motor drive assembly can be varied electronically to achieve similar function as a mechanical differential component.
e) This configuration represents in-wheel drivetrain. This configuration results in further reduction of powertrain weight. The motor and fixed gearing in this configuration are integral to the driving wheel. For this system, the planetary gearbox is mostly adopted due to high reduction gear ratio and inline arrangement requirements.
f) Further simplification of the in-wheel drive is possible. This configuration eliminates the fixed gearing by adopting a low speed in-wheel electric motor. As such, vehicle speed control is tantamount to control of motor speed.

5.1.1.2 Hybrid Electric Vehicle (HEV)

HEV is a vehicle with at least two energy sources of which one produces electrical power [2]. This is a broad definition encompassing a variety of possible combination of energy sources. Herewith, the use of the term HEV is limited to only ICE and battery energy source combinations. Figure 5.2 shows typical configurations of HEV powertrain as series hybrid, parallel hybrid, series-parallel hybrid and complex hybrid.

a) *Series Hybrid*
In this configuration, the ICE is coupled to a generator which produces electricity to either charge the battery or power the electric motor directly for electric propulsion. The series hybrid powertrain configuration is akin to an ICE-assisted EV range extender. An advantage of this configuration is the flexibility with ICE and generator placement. However, its

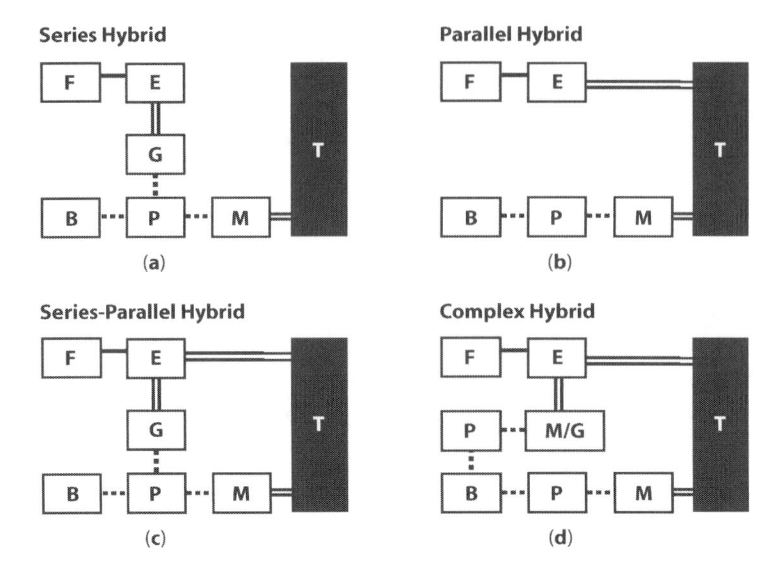

Figure 5.2 Typical HEV powertrain configuration [2].

downside is the high cost of this configuration due to the fact that the entire system, ICE and electric motor, has to be sized for maximum power output.

b) *Parallel Hybrid*

In this configuration, both the ICE and electric motor are coupled directly to the driveshaft to provide propulsion. Herewith, the vehicle can be propelled by either ICE alone, electric motor alone or both. The parallel hybrid configuration can be viewed as an electric-assisted ICE vehicle, with an added advantage of lower emissions and fuel consumption. The advantage of a parallel hybrid configuration over the series is that only the ICE needs to be rated for maximum power output.

c) *Series-Parallel Hybrid*

This configuration is a fusion of both the series hybrid and parallel hybrid powertrain configuration and benefits from powering option flexibility. However this flexibility comes with control system complexity and increased system configuration cost.

d) Complex Hybrid
As hybrid powertrain evolves, the complex hybrid configuration accounts for such concepts that do not fit into the series hybrid, parallel hybrid and series-parallel hybrid configurations. Figure 5.2(d) shows an example of a complex hybrid powertrain configuration with a bi-directional energy flow between the ICE and battery-electric motor powering option.

5.1.1.3 Fuel Cell Electric Vehicle (FCEV)

FCEV has fuel cells as the primary energy source, and the electric motor to provide propulsion. The fuel cell can be fuelled by hydrogen or methanol. As fuel cells do not accept regenerative energy, batteries are usually included as an auxiliary energy source.

5.1.2 EV Powertrain Technologies

5.1.2.1 Energy Storage System

Energy storage such as batteries are a vital part of the electric vehicle powertrain. The following energy storage systems are commonly used in an EV.

1) Lithium-ion (Li-ion) Battery
Li-ion batteries are widely adopted for EVs. They are capable of having a very high voltage and charge storage per unit mass and unit volume. Li-ion batteries can use a number of different materials as electrodes, such as lithium cobalt oxide for cathode and graphite for anode, although for portable electronic devices, e.g., laptop, mobile phone, etc. For EVs, several cathode material options are possible, such as lithium manganese oxide, lithium iron phosphate, lithium titanate, amongst others. When compared to other rechargeable battery chemistry such as Nickel-Cadmium (Ni-Cd) and Nickel Metal Hydride (Ni-MH), Li-ion has the following advantages [4]: high energy density (~100-265 Wh/kg) as shown in Figure 5.3, high specific power, broad useable temperature range, low maintenance (i.e., no need for scheduled cycling to maintain battery life), no voltage depression (i.e., a memory effect artefact of a battery remembering low battery capacity due to repeated partial discharge and charging evident in Ni-Cd and Ni MH), and low self-discharge. However, their limitations are overheating, issues with ageing, and relatively high cost.

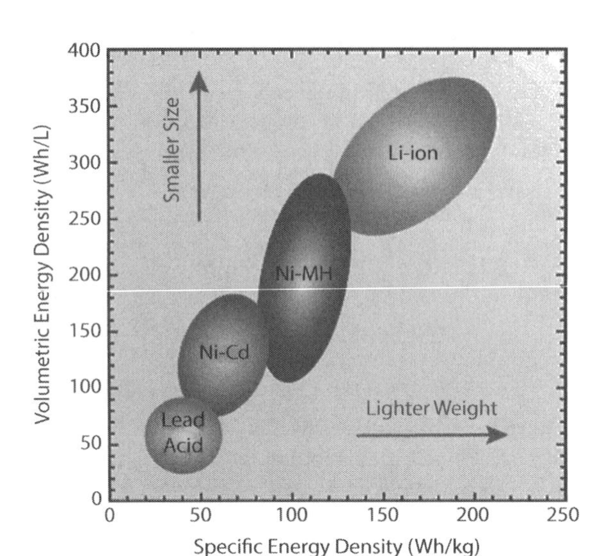

Figure 5.3 Comparison of battery chemistry energy density [4].

2) *Nickel Metal Hydride (Ni-MH) Battery*
These are used mostly in HEV. They have a reasonable specific power and specific energy, long useful life, and are tolerant to abuse conditions. Their main limitations are high cost, high self-discharge, poor high temperature performance, and the need to control hydrogen loss.
3) *Lead Acid Battery*
These batteries are mostly used in EVs to power auxiliary systems. They are inexpensive, they have a high specific power and are tolerant to abuse conditions. The challenges with these batteries are poor cold temperature performance, low specific energy, and poor cycle life.
4) *Ultra-Capacitor*
The operation of an ultra-capacitor is such that energy is stored in a polarised liquid between an electrode and electrolyte. They are used in EVs as they provide additional power for vehicles during acceleration and hill climbing as well as in recovering braking energy. They may also be useful as secondary energy-storage devices in electric-drive vehicles because they help electrochemical batteries level load power.

5.1.2.2 *Electric Motor*

Electric motors are the main motive component in an EV powertrain. Together with the battery, the electric motor transforms electrical energy to rotational or mechanical energy transmitted to the vehicle wheels for propulsion. Although there are various types of electric motors based on applications, the common electric motor technology applied in EV is shown in Figure 5.4 [3].

1) *Direct Current Motor*
   DC motor technology is becoming phased out for EV application due to a key drawback of high maintenance arising from its carbon brushes/commutators needing electrical contact and over time, wearing off. Apart from these, they can provide high starting torque, handle sudden increase in load, are low cost, have simple construction and are easy to control.
2) *Brushless Direct Current Motor*
   Brushless DC motor overcomes the inherent weakness of DC motor, the need for electrical contact via carbon brushes/commutator. It does away with the commutator by having the windings on the stator and the permanent magnet on the rotor; as such, there is no need for carbon brushes. It also can provide high starting torque and handle sudden increase in load. It is also low cost, with a simple construction, easy to control and low maintenance.
3) *Alternating Current (AC) Induction Motor*
   AC induction motors, also referred to as asynchronous induction motors, are used in EV due to their high efficiency, good speed control and low maintenance due to the absence of a commutator. They occur in construction as either squirrel cage or wound-type AC induction motors. The operation of this motor is such that a current through

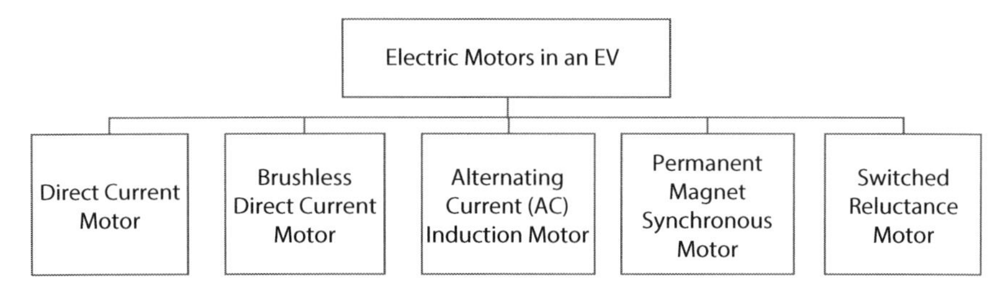

Figure 5.4 Types of electric motor in an EV.

rotor windings induces a magnetic field, which can interact with the magnetic field of the stator to produce a force on the rotor, causing rotation.

4) *Permanent Magnet Synchronous Motor*
   Permanent magnet synchronous motors (PMSM) are compact, have a high power-to-weight ratio, provide high starting toque, are very stable in operation, and are low maintenance. As such, they are widely used in various EVs such as light vehicles and buses. They occur in construction as either surface PMSM or internal PMSM. The operation of this motor is such that the interaction of the rotating magnetic field of the stator and the constant magnetic field of the rotor formed from permanent magnet located in the rotor (i.e., surface or internal construction), produce mechanical rotation of the rotor. The key difference between a PMSM and an AC induction motor is the rotor construction.
5) *Switched Reluctance Motor*
   Switched reluctance motors are gaining adoption in EVs due to their high efficiency, reliability, wide speed range, good controllability, and simple construction. Based on construction, the following configurations are applied, such as singly salient and doubly salient constructions. The operation of this motor is such that the rotor has salient poles and a cage so that it starts like an induction motor and runs like a synchronous motor.

5.1.2.3 Power Electronics

The power electronics in EVs as shown in Figure 5.5 consist of converter/inverter components and protection components. In general, they are responsible for power exchanges between EV powertrain sub-systems, such as battery, battery management system (BMS), electric motor, and electric accessories. Converter/inverter components enable electrical power conversion by periodically fast switching available power source on and off with the use of semiconductor devices such as field-effect transistors (FETs), insulated gate bipolar transistors (IGBTs), etc., acting as switches, and energy storage devices, e.g., inductors and capacitors used to store energy to filter the sharp-edged waveforms created by the fast switching operation [3]. On the other hand, protection components such as transformers are used to provide isolation, safety, and voltage/current conversions.

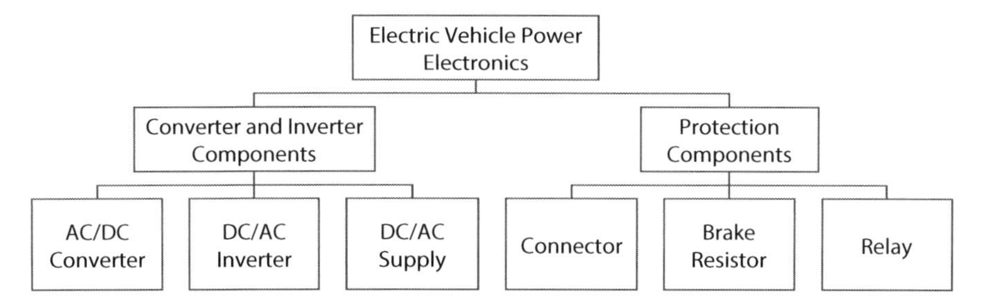

Figure 5.5 Classification of electric vehicle power electronics.

5.2 Battery Fault Diagnosis

Failures in the battery as an energy source for EVs can have a detrimental economic effect as well as an impact on its environmental sustainability. At the beginning, battery fault detection focused on monitoring overheating, overcurrent and overvoltage; however, with advances in battery technology, fault detection and isolation techniques have moved on.

Generally, faults in batteries can be categories as follows [5]:

1) *Battery Temperature Fault:* There is a minimum and maximum working temperature range for batteries (5 °C – 45 °C) [6]. When a battery operates out of this temperature range it produces undertemperature fault and overtemperature fault.
2) *Battery Current Fault:* This fault occurs in batteries due to the presence of a short circuit.
3) *Battery Voltage Fault:* In a battery, this fault manifests as low voltage fault, overvoltage fault, high voltage loop fault, and high voltage insulation fault.
4) *Battery State of Charge (SOC) Fault:* This is a fault in battery charging. It manifests as pre-charge fault, overcharging fault and overdischarging fault.

5.2.1 Battery Management System (BMS)

A BMS is an electronic control circuit that monitors and regulates battery charging and discharging. As shown in Figure 5.6, the BMS must be able to satisfy the following critical functions [7]:

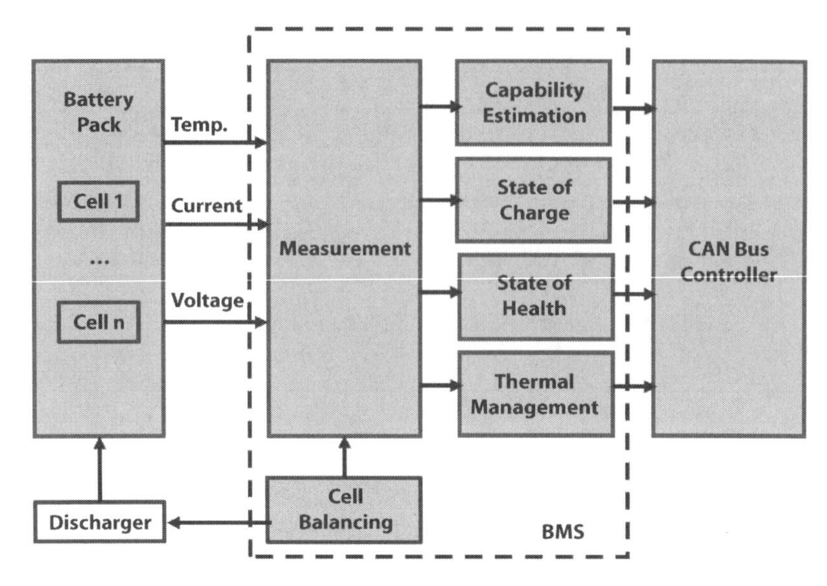

Figure 5.6 Schematic of battery management system [7].

1) *Battery Protection*
   It is important for EV batteries to operate within predefined safe limits to ensure the safety of the user as well as the vehicle [6]. The BMS regularly monitors the battery parameters (temperature, current and voltage) to ensure they are within safe working limits (i.e., cell voltage at full charge, 4 V, and, cell voltage at full discharge, 2 V). To prevent overheating in a Li-ion battery due to overcharging, the BMS would monitor both the cell and pack voltages to control the supply current. Another protective feature of the BMS is that it ensures complete isolation of the battery pack high voltage from the chassis.
2) *Battery Monitoring*
   The BMS estimates the state of charge (SOC) and state of health (SOH) using battery parameters (temperature, current and voltage). SOC refers to the available energy in the battery, and indicates a vehicle's range on a single charge. On

the other hand, SOH refers to a battery's current condition as compared to its original state. SOH also indicates a battery's suitability for application. Additionally, BMS monitors for anomalies in battery parameters (temperature, current and voltage) and trigger fail-safe or corrective actions to protect the health of the battery.

3) *Battery Optimisation*

Batteries such as Li-ion operate optimally when SOC is maintained in practical applications within minimum 15% SOC and maximum 50% SOC battery charge profile [8], although a battery can achieve 0% SOC, which is fully discharged, and 100% SOC, which is fully charged. This is important, as overcharging and overdischarging causes degradation of battery charge capacity and duration of useable life. As such, during charging, the BMS estimates the safe charge input current and communicates such to the on-board charger during AC charging or electric vehicle supply equipment (EVSE) for DC charging. Also, during discharging, the BMS prevent very low cell voltage and communicates continuously with the motor controller to avoid this. Another optimisation function of the BMS is cell balancing. Individual cells in the battery pack over time can develop differences in capacity (i.e., cell voltage); this worsens with each discharge and charge cycle. This imbalance in cell voltage limits the amount of energy that can be discharged from the battery, and also how much the battery pack can be charged. Hence, cell balancing ensures battery cells are maintained at equal voltage levels and optimises charge capacity utilisation of the battery pack.

4) *Communication*

BMS communicates with other electronic communication units (ECU) in the vehicle to ensure a fault-free operation such as communicating battery parameters to the motor controller during discharging or during charging, communicating safe input current with the on-board charger (AC charging) or electric vehicle supply equipment (DC charging).

5.2.2 Model-Based FDI Approach

5.2.2.1 Battery Modelling

A battery represents a nonlinear thermal-electrochemical system. Battery models help answer the following questions: what is the battery state of charge, state of health, and performance and lifespan trade-offs due to chemical and mechanical degradation of the battery caused by environment and cycling? [9, 10]. Two approaches are generally applied for battery modelling [11]: equivalent circuit model and electrochemical model. The equivalent circuit battery model refers to a simplified theoretical circuit model that retains all the electrical properties of the battery. Several such models have been proposed [11–13]: Rint battery model, Thevenin battery model, PNGV battery model [14], Dual Polarisation battery model and RC battery model. On the other hand, the complex electrochemical model [15] would consider the battery cell electrical dynamics at the molecular level.

1) *Rint Battery Model*
   The Rint battery model, as shown in Figure 5.7, considers the battery open-circuit voltage as an ideal voltage source and the battery internal resistance as a resistor [11, 12]. Both parameters, open-circuit voltage and internal resistance, are a function of SOC, SOH and temperature. I_L is load current with a positive value at discharging and a negative value at charging,
   Applying circuit theory to Figure 5.7, the following governing equation applies [12]:

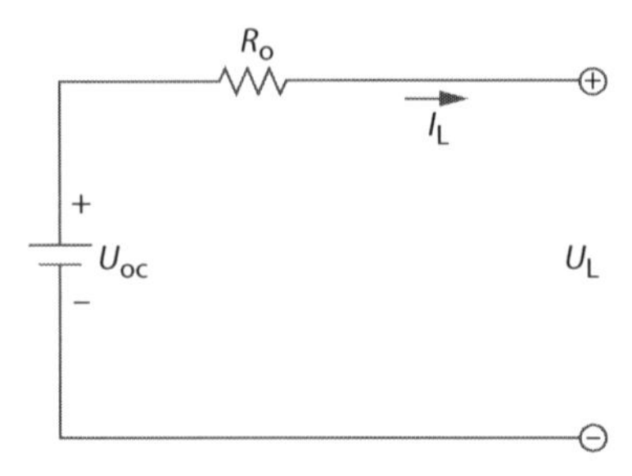

Figure 5.7 Rint battery model [11, 12].

$$U_L = U_{OC} - I_L R_O \tag{5.1}$$

where U_{oc} is battery open-circuit voltage, R_o is battery internal resistance, U_L is battery terminal voltage and I_L is load current.

2) *Thevenin Battery Model*

The Thevenin battery model is an extension of the Rint battery model. It addresses the transient effect on the battery due to charging and discharging. This is implemented in the model by incorporating a parallel RC network in series with the Rint battery model as shown in Figure 5.8. The model consists of the open-circuit voltage U_{oc}, battery internal resistances (ohmic resistance R_o and polarisation resistance R_{th}) and equivalent battery capacitance C_{th}.

Applying circuit theory to Figure 5.8, the following governing equation applies [12]:

$$\dot{U}_{TH} = -\frac{U_{TH}}{R_{TH}C_{TH}} + \frac{I_L}{C_{TH}} \tag{5.2}$$

$$U_L = U_{OC} - U_{TH} - I_L R_o \tag{5.3}$$

where U_{oc} is battery open-circuit voltage, U_L is battery terminal voltage and I_L is load current.

3) *PNGV Battery Model*

The PNGV battery model as shown in Figure 5.9 is an extension of the Thevenin battery model. It addresses the

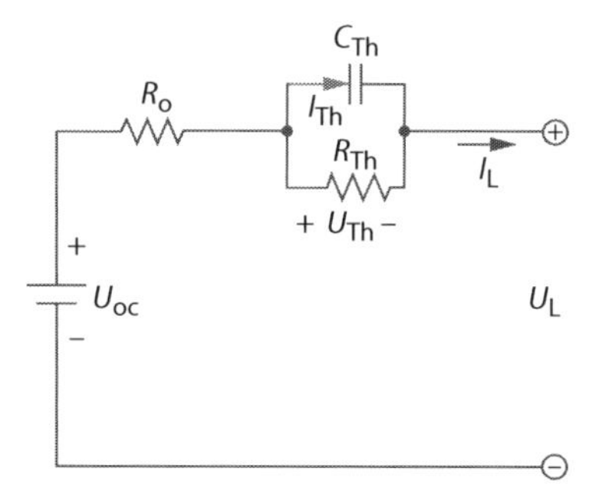

Figure 5.8 Thevenin battery model [12].

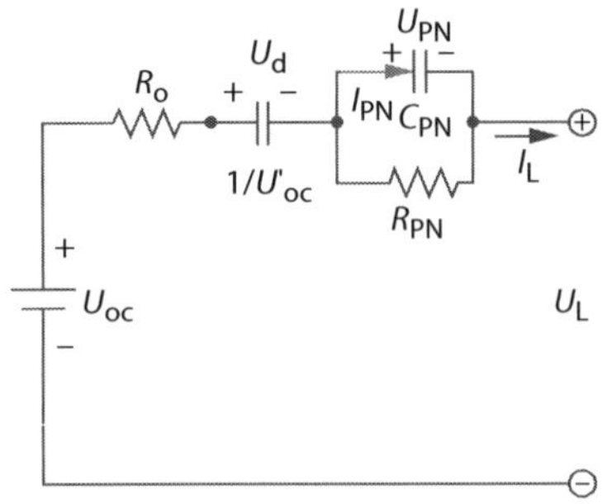

Figure 5.9 PNGV battery model.

changing open-circuit battery voltage due to time-related accumulation of load current. This is implemented by incorporating a capacitor in series to the Thevenin battery model.

Applying circuit theory to Figure 5.9, the following governing equation applies [12]:

$$\dot{U}_d = U'_{OC} I_L \quad (5.4)$$

$$\dot{U}_{PN} = \frac{U_{PN}}{R_{PN}C_{PN}} + \frac{I_L}{C_{PN}} \quad (5.5)$$

$$U_L = U_{OC} - U_d - U_{PN} - I_L R_o \quad (5.6)$$

where U_{oc} is battery open-circuit voltage, U_L is battery terminal voltage, I_L is load current, battery internal resistances (ohmic resistance R_o and polarisation resistance R_{PN}), U_d is the voltage across series capacitor $1/U'_{oc}$ and U_{PN} is voltage across polarisation capacitor C_{PN} and I_{PN} is the output current from C_{PN}.

4) *Dual Polarisation (DP) Battery Model*

The DP battery model extends the Thevenin battery model by considering the effect of concentration polarisation as well as the electrochemical polarisation. This is implemented by incorporating an RC parallel network in

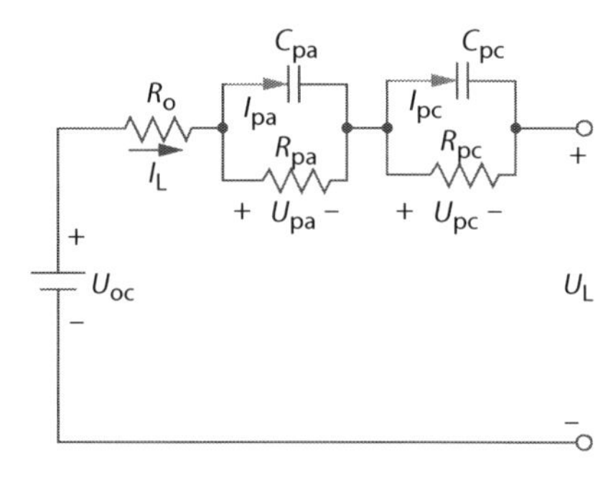

Figure 5.10 DP battery model [12].

series with the Thevenin battery model. The DP battery model as shown in Figure 5.10 consists of the open-circuit battery voltage U_{oc}, battery ohmic resistance R_{o}, battery polarisation resistances (i.e., effective electro-chemical polarisation resistance R_{pa} and effective concentration polarisation resistance R_{pc}), and polarisation capacitance (i.e., electro-chemical polarisation capacitance C_{pa} which accounts for charging and discharging transient effects and the concentration polarisation capacitance C_{pc}).

$$\dot{U}_{pa} = -\frac{U_{pa}}{R_{pa}C_{pa}} + \frac{I_L}{C_{pa}} \tag{5.7}$$

$$\dot{U}_{pc} = \frac{U_{pc}}{R_{pc}C_{pc}} + \frac{I_L}{C_{pc}} \tag{5.8}$$

$$U_L = U_{oc} - U_{pa} - U_{pc} - I_L R_o \tag{5.9}$$

where $\dot{U}_{pa}$ and $\dot{U}_{pc}$ are derivatives of the voltage across the electro-chemical polarisation capacitor U_{pa}, and concentration capacitor U_{pc} respectively, open-circuit voltage U_{oc}, and I_L is load current.

5) *RC Battery Model*

The RC battery model as shown in Figure 5.11 considers a battery as two capacitor (surface capacitor C_c and bulk capacitor C_b), and three resistors (terminal resistor R_t, end resistor

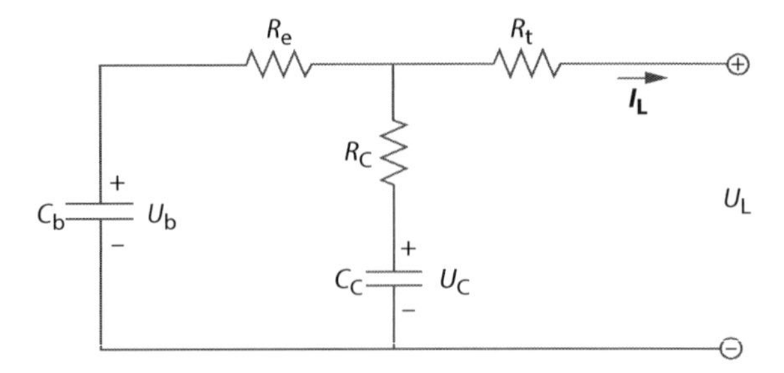

Figure 5.11 RC battery model [11, 12].

R_e, capacitor resistor R_c) [11, 12]. Here, surface capacitor C_c represents the small capacitive surface effect of the battery, and the bulk capacitor C_b represents battery charge storage.

Applying circuit theory to Figure 5.11, the following governing equation applies [11, 12]:

$$\begin{bmatrix} \dot{U}_b \\ \dot{U}_c \end{bmatrix} = \begin{bmatrix} \dfrac{-1}{C_b(R_e+R_c)} & \dfrac{1}{C_b(R_e+R_c)} \\ \dfrac{1}{C_c(R_e+R_c)} & \dfrac{-1}{C_c(R_e+R_c)} \end{bmatrix} \begin{bmatrix} U_b \\ U_c \end{bmatrix} + \begin{bmatrix} \dfrac{-R_c}{C_b(R_e+R_c)} \\ \dfrac{-R_e}{C_b(R_e+R_c)} \end{bmatrix} \begin{bmatrix} I_L \end{bmatrix} \tag{5.10}$$

$$\begin{bmatrix} U_L \end{bmatrix} = \begin{bmatrix} \dfrac{R_c}{(R_e+R_c)} & \dfrac{R_e}{(R_e+R_c)} \end{bmatrix} \begin{bmatrix} U_b \\ U_c \end{bmatrix} + \begin{bmatrix} -R_t - \dfrac{R_c R_e}{(R_e+R_c)} \end{bmatrix} \begin{bmatrix} I_L \end{bmatrix} \tag{5.11}$$

where $\dot{U}_b$ and $\dot{U}_c$ are derivatives of the bulk capacitor voltage U_b and surface capacitor voltage U_c, and I_L is load current.

6) *Thermal-Electrochemical Battery Model*

Thermal-electrochemical battery model as shown in Figure 5.12 couples the battery cell thermal energy equation with the multiphase micro-macroscopic electrochemical model via the

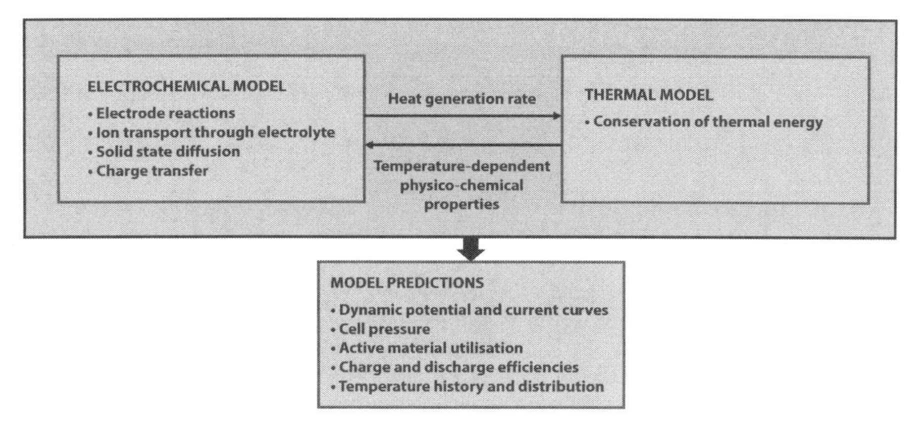

Figure 5.12 Schematic of thermal-electrochemical [15].

heat generation and temperature-dependent physicochemical properties such as diffusion coefficient and ionic conductivity of the electrolyte [15, 16]. Electrochemical battery models are capable of predicting both the spatial and average temperature in the battery cell, cell charge and discharge efficiencies, cell pressures, dynamic potential and current curves, and active material utilisation [9, 15, 16].

5.2.3 Signal Processing-Based FDI Approach

Although the BMS using measurement of battery voltage, current and temperature can estimate the batteries' state of charge and health, limitations exist regarding individual fault type diagnostic and accuracy in state estimation [5]. For this application, Figure 5.13 shows an advanced battery diagnostics system using signal processing-based. This system integrates both hardware components (i.e., signal modulation circuit and data acquisition card), as well as software components (i.e., acquisition module and management module). The hardware components monitor the battery parameters (i.e., voltage, current and temperature) and modulate the signals for storage by the acquisition card. Within the acquisition and management modules, software applies a signal processing-based approach on the stored data to perform performance and diagnostic checks. Some of these techniques are presented below.

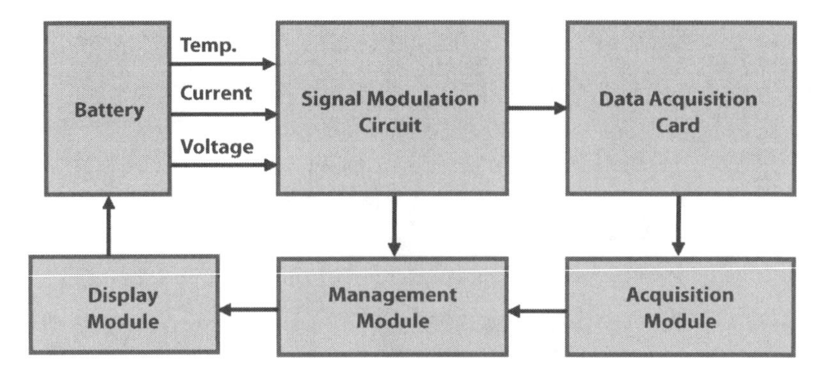

Figure 5.13 Schematic of advanced battery diagnostic system [5].

5.2.3.1 *State of Charge (SOC) Estimation*

Battery SOC represents its remaining charge capacity. Mathematically, SOC is a ratio between the battery's current capacity to its nominal capacity [17]. It is an important parameter required by the battery management system for effective control of the battery. Chang [17] reviewed various approaches for SOC estimation. Broadly, there are four categories of SOC estimation methods: direct measurement methods, book keeping estimation, adaptive systems, and hybrid methods. In the adaptive systems and hybrid methods, signal processing-based techniques such as Kalman filter [18], extended Kalman filter [19, 20] and unscented Kalman filter [21, 22] are important and provide improvements in accuracy.

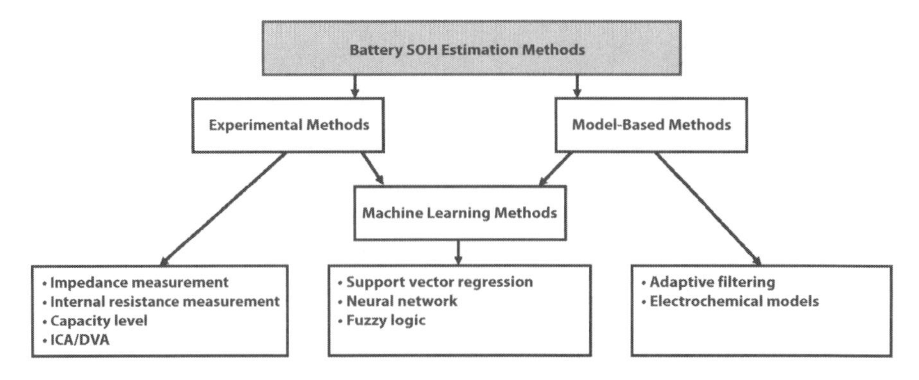

Figure 5.14 Overview of battery SOH estimation methods [23].

5.2.3.2 *State of Health Estimation*

Battery SOH is an indication of the ability of a battery to store and release a charge as compared to a new battery. Monitoring the SOH of a battery is important for prolonging the battery lifetime. Noura [23] provides a review of methods for estimating the SOH of batteries. As shown in Figure 5.14, signal processing-based techniques such as Kalman filter, extended Kalman filter and unscented Kalman filter as applied in the adaptive systems approach are relevant for accurate battery SOH estimation [23–25].

5.3 Electric Motor Fault Diagnosis

5.3.1 Electric Motor Faults

Although electric motor types are varied in construction, they all have the following components: stator, rotor, and bearings as shown in Figure 5.15. As such, a common trend can be observed regarding fault occurrence.

Generally, faults in an electric motor can be grouped as [26].

5.3.1.1 *Mechanical Fault*

These are faults that affect the integrity of the bearing and the eccentricity of the rotor in the stator. Bearing-related fault can manifest as inner and outer bearing race defects, rolling-element defect and cage defect. Eccentricity-related faults can be due to bend shaft, air gap irregularity, etc.

5.3.1.2 *Electrical Fault*

Electrical fault is associated with faults in the rotor and stator. Rotor-related fault can manifest as broken rotor bar, short circuit rotor winding, cracked

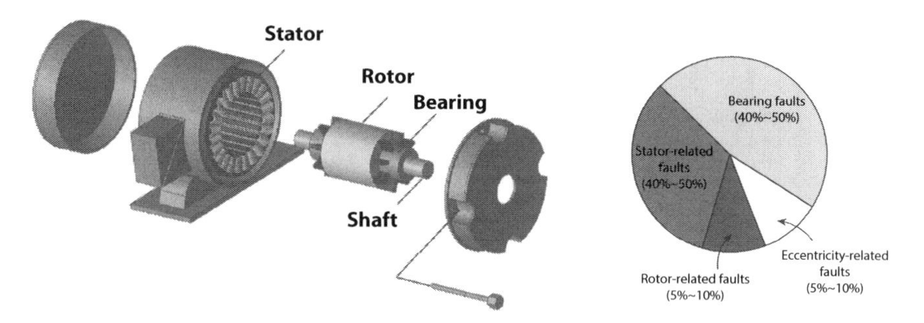

Figure 5.15 Parts of an electric motor and failure mode [5].

rotor end ring, etc. Stator-related fault can occur as open circuit stator winding, short circuit stator winding, and abnormal connection of stator winding.

5.3.2 Signal Processing-Based FDI Approach

5.3.2.1 Motor Current Signature Analysis (MSCA)

MSCA is a fault detection and isolation technique applicable for induction motors [27]. The concept has its origin dating back to 1970 in the nuclear industry where it was applied for monitoring the health of electric motors which were located in inaccessible and often hazardous conditions [27]. In recent times, MSCA is also applied to EVs and other industries to monitor the health of electric motors [28–30]. MSCA is quite advantageous, as it does not interrupt the operation of the electric motors. As shown in Figure 5.16, MSCA acquires the current signal (i.e., stator current) from one phase of the motor supply without interrupting its operation. In MCSA, the current signal is processed using a signal-processing technique such as Fast Fourier Transform (FFT) to obtain the frequency spectrum also called current spectrum. MCSA uses the current spectrum of the electric motor for locating fault frequencies. As shown in Figure 5.17, the current spectrum of the electric motor with a fault is different from the current spectrum of a healthy motor.

With MSCA, the following faults in an electric motor are detectable [27]:

1) *Air Gap Eccentricity*
   Air gap eccentricity occurs when the air gap between the rotor and stator in the electric motor is not uniform. This can

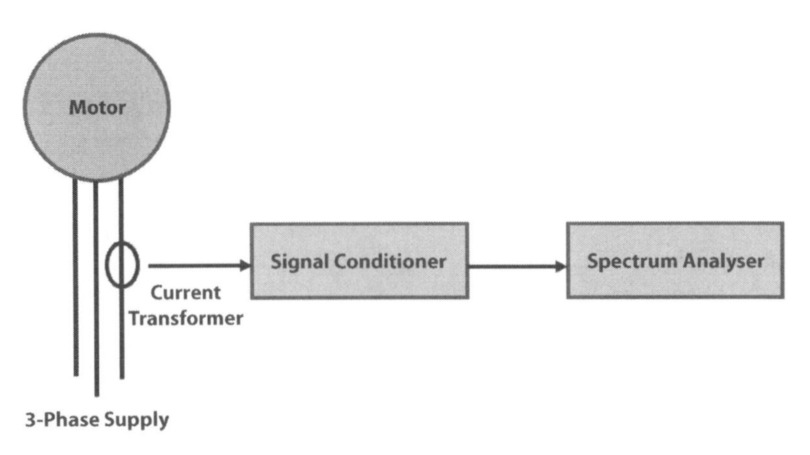

Figure 5.16 Overview of MSCA approach [27].

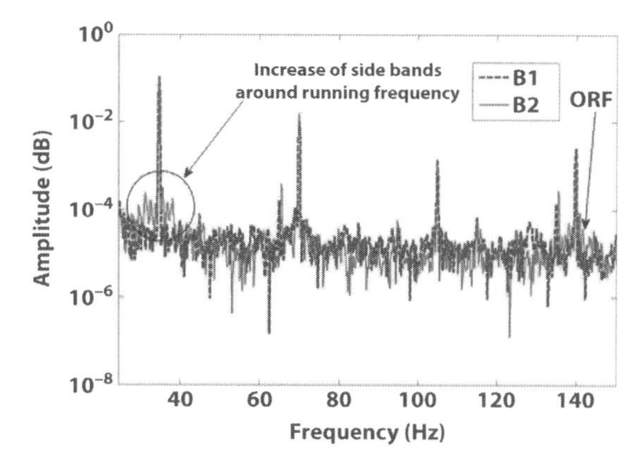

Figure 5.17 Current spectrum from MSCA of a healthy electric motor (B1) and faulty electric motor with damaged bearings (B2) [31].

occur due to assembly and manufacturing errors. This can take the form of the following:

- Static Eccentricity: Occurs when the rotor geometrical and rotational centres are identical; however, they are different from the stator geometrical centre. This can happen due to manufacturing tolerances not being met.
- Dynamic Eccentricity: Occurs when the rotor geometrical centre does not coincide with its rotational centre.
- Mixed Eccentricity: This is a combination of both the static and dynamic eccentricities.

In determination of static eccentricity using MSCA, sidebands would occur around the eccentricity frequency f_{ec}, determined by:

$$f_{ec} = f_g \left\{ (R \pm n_d) \left(\frac{1-s}{p} \right) \pm n_{ws} \right\} \tag{5.12}$$

where f_{ec} is eccentricity frequency, f_g is line or supply frequency, R is number of rotor bar, s is slip, p is number of pole pairs, $n_d = \pm 1$ and $n_{ws} = 1,3,5,\cdots$

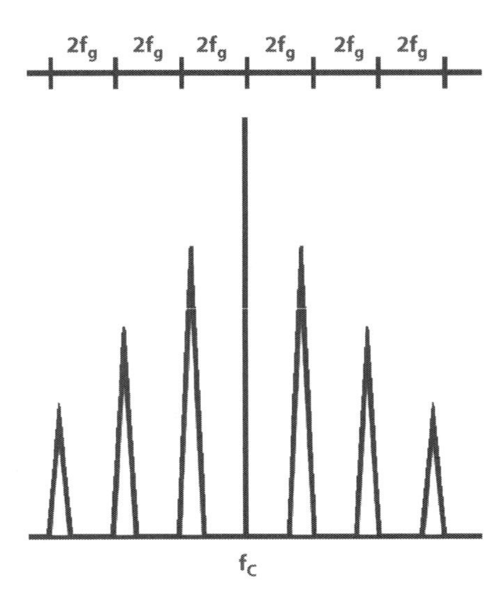

Figure 5.18 Air gap static eccentricity current spectrum [27].

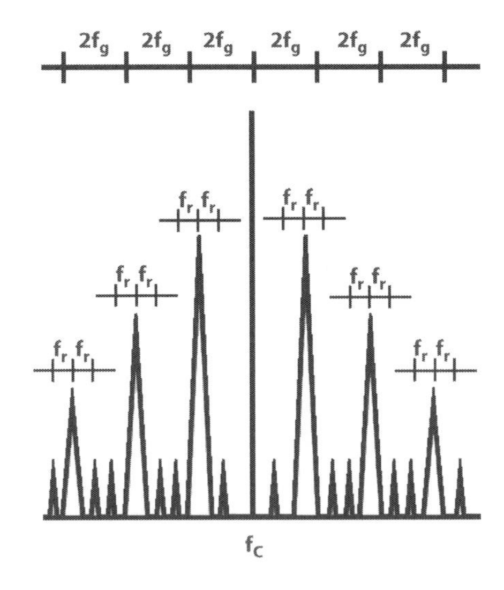

Figure 5.19 Air gap dynamic eccentricity current spectrum [27].

Slip (s) is determined by:

$$s = \frac{N_s - N_r}{N_s} \tag{5.13}$$

where N_s is rotor speed and N_r is rotor synchronous speed.

Also, static eccentricity can occur as sideband around the centre frequency f_c given below and also shown in Figure 5.18.

$$f_c = Rf_g \tag{5.14}$$

where f_g is supply frequency, and R is number of rotor bar. Furthermore, the occurrence of air gap dynamic eccentricity is detected by further harmonics to the current spectrum as shown in Figure 5.19. These new harmonics are modulated by the mechanical rotor frequency f_g.

2) *Broken Rotor Bar*
Broken rotor bars can occur in induction motors due to arduous duty cycles. In themselves, they would not cause failure of the induction motor; however, they can create secondary damage, e.g., broken rotor bar parts can damage the winding. With MCSA, broken rotor bar fault can be detected in the current spectrum by the presence of the broken rotor bar frequency component, determined by [27]:

$$f_{brb} = f_g\left[k\left(\frac{1-s}{p}\right) \pm s\right] \tag{5.15}$$

where f_{brb} is broken rotor bar frequency, f_g is line or supply frequency, p is pole pairs, s is slip, and $k = 1,2,3\cdots$.

3. *Bearing Fault*
Bearing damage in electric motors is primarily caused by misalignment especially during installation. With MSCA, bearing faults are detected by the presence of bearing fault frequency components f_0 and f_1 in the current spectrum. For bearings with 6-12 rolling elements, these frequency components can be determined as follows [27]:

$$f_o = 0.4nf_{rm} \tag{5.16}$$

$$f_1 = 0.6nf_{rm} \tag{5.17}$$

where is lower frequency, is upper frequency, n is number of rolling elements, and is the mechanical rotational frequency of the rotor.

4) *Shorted Turns-In Stator Windings*
Shorted stator winding fault is a very common fault in induction motors. The stator current can be continuously monitored in order to predict incipient fault in the stator windings. In MSCA, the shorted turns-in stator windings fault can be detected by the presence of shorted stator windings frequency component determined by [27]:

$$f_{st} = f_g\left[\frac{n}{p}(1-s)\pm k\right] \tag{5.18}$$

where f_{st} is shorted stator winding frequency, f_g is line or supply frequency, n is number of rolling elements, p is pole pairs, s is slip and k = 1,2,3 …

5) *Load Effects*
During the operation of an electric motor, as a result of duty cycle demand, there can be variation or fluctuation in the transmitted torque. This load effect can also be monitored by MSCA. Using MSCA, variability in torque can be detected by the presence of the following frequency component in the current spectrum, determined as such [27]:

$$f_{load} = f_s \pm mf_r = f_s\left[1\pm m\left(\frac{1-s}{p}\right)\right] \tag{5.19}$$

where f_s is line or supply frequency, f_r is mechanical rotational frequency, p is number of pole pairs, s is slip and km = 1,2,3 …

5.4 Power Electronics Fault Diagnosis

The reliability of power converter/inverter components depends primarily on the endurance of its main component, the power switches [32]. There are varied types of power switches such as thyristor, gate turn-off thyristor (GTO), power bipolar-junction transistor (BJT), power metal-oxide field effect transistor (MOSFET), insulated gate bipolar transistor (IGBT), static induction transistor (SIT), static induction thyristor (SITH), MOS controlled thyristor (MCT), and MOS turn-off thyristor (MTO); for EV power converter/inverter, IGBT is the widely adopted switching technology [3, 32, 33].

For power switching devices, the most common types of faults are [32]:

1) Open switch fault
2) Short switch fault

5.4.1 Signal Processing-Based FDI Approach

5.4.1.1 Open Switch Fault

An open switch fault in an IGBT can be due to a break of its bond wires or fault in the gate drive. This results in a DC offset in the faulty phase/leg. Figure 5.20 shows an example of an open switch fault in S_{ap} a three-phase two-leg inverter controlling an AC motor. As the IGBT with the open switch fault would not turn ON, in the case of motor operation, the current in that phase becomes zero for a half-cycle, either positive or negative half-cycle depending on whether it is upper IGBT or lower IGBT. This behaviour results in the DC offset observed in practice from the faulty phase.

To detect open switch fault in power drives, several approaches are reported in literature [5, 32, 34–37]; however, two broad methods are evident [32]: voltage sensor measurements [35] and software defined-techniques [36, 37].

Voltage sensor measurement techniques are fast acting and require short detection time. However, they require additional voltage sensor to insert into the system. Depending on the location of insertion of the voltage sensor, the following voltage measurement indicators are relevant for power drive open switch fault diagnosis [32, 35]:

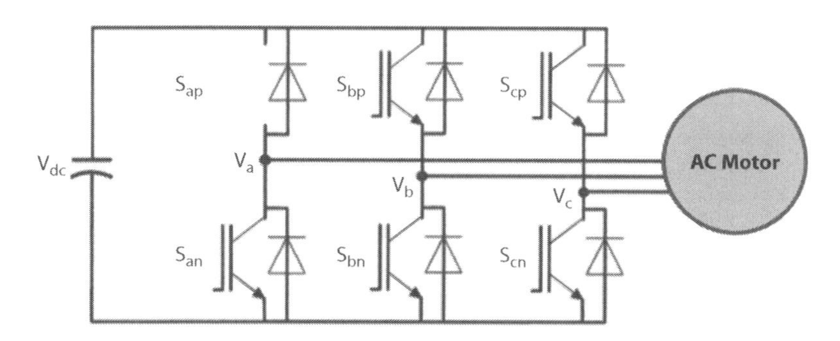

Figure 5.20 Three-phase two-level inverter with IGBT open switch fault at S [32].

- Inverter pole voltage measurement
- Machine phase voltage measurement
- System line voltage measurement
- Machine neutral voltage measurement

Software defined-techniques do not require additional voltage sensors, although they require a longer detection time. In this category, various techniques for tracking the current vector trajectory [36, 38] and 3-phase current mean value [39] as indicators for open switch fault in power drives have been proposed. Table 5.2 provides an overview of techniques used for detecting open switch fault in power drives [5].

Table 5.2 Summary of power drive open switch fault detection [5].

Fault	Symptoms	Fault indicator	Approach	Comments
Open switch fault	DC current offset	Current vector trajectory	Park's method	Load dependent
			Normalised DC current method	Complicated
			Modified normalised DC current method	High efficient
		3-phase current mean value	Wavelet-neural network method	High cost, intelligent
			Current deviation method	High implementation effort
		Voltage	Voltage comparison in time domain	Extra sensor needed
			Voltage sensing by lower switch	Quick detection

5.4.1.2 Short Switch Fault

A short switch fault in an IGBT may be due to the malfunctioning of the gate drive or permanent damage to the IGBT. This results in practice to a non-zero DC current in the shortened phase/leg. Short switch fault results very quickly in the failure of the e-drive, as such isolating the shortened phase with approach such as a fast-acting fuse or field programmable gate array, etc., is required especially for fault tolerant mode of operation. Figure 5.21 shows an example of a short switch fault in Sap a three-phase two-leg inverter controlling an AC motor. When the IGBT detects the short circuit fault, all the IGBTs are turned OFF by hardware protection. In the event of IGBT permanent damage, its corresponding phase is permanently connected to the dc link positive bus or negative bus depending on upper IGBT or lower IGBT is damaged. Provided the machine is operational, current flows through the shorted IGBT and remaining freewheeling diodes of the inverter.

To detect short switch fault in power drives, the following current and voltage measurement parameters are relevant for fault diagnosis [5]:

- Phase current
- Device current
- Gate voltage

Table 5.3 highlight various techniques for detecting short switch circuit.

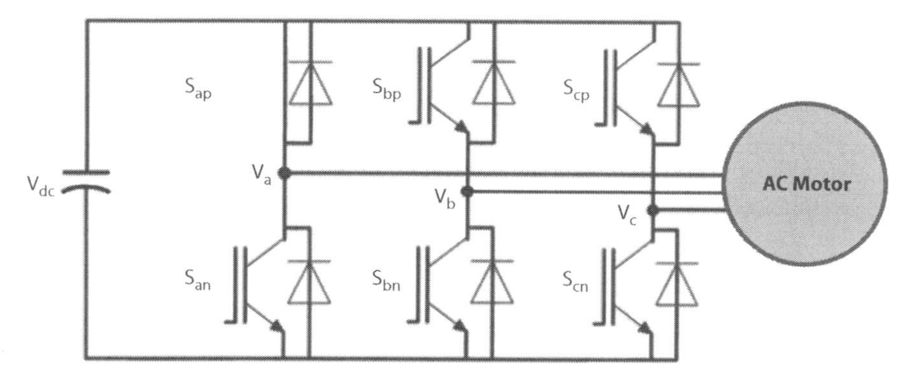

Figure 5.21 Three-phase two-level inverter with IGBT short switch fault at S_{ap} [32].

Table 5.3 Summary of power drive short switch fault diagnosis [5].

Fault	Symptoms	Fault indicator	Approach	Comments
Short switch fault	Non-zero DC current component	Phase current	Average current Park's vector approach	No protection
		Device current	Current mirror method	High cost
		Gate Voltage	Protection by gate voltage limiting	Inaccurate detection
			Gate voltage sensing method	Complicated
			Two-step gate pulse	Reliable
			Voltage and time criterion	Fast reaction, fault tolerant

5.5 Conclusions

In this chapter, an overview of EV powertrain technologies and various configurations has been presented. Recent research on fault detection for powertrain components such as battery, electric motor and power electronics has also been discussed, with emphasis on the application of various model-based and signal processing-based fault detection and isolation approaches relevant for EVs.

References

1. Un-Noor, F., Padmanaban, S., Mihet-Popa, L., Mollah, M.N., and Hossain, E. (2017) A Comprehensive Study of Key Electric Vehicle (EV) Components,

Technologies, Challenges, Impacts, and Future Direction of Development. *Energies*, **10** (8).
2. Chan, C.C. (2002) The state of the art of electric and hybrid vehicles. *Proc. IEEE*, **90** (2), 247–275.
3. Chau, K., and Wang, Z. (2005) Overview of power electronic drives for electric vehicles. *HAIT J. Sci. Eng. B*, **2** (5–6), 737–761.
4. Landi, B.J., Ganter, M.J., Cress, C.D., DiLeo, R.A., and Raffaelle, R.P. (2009) Carbon nanotubes for lithium ion batteries. *Energy Environ. Sci.*, **2** (6), 638–654.
5. Lin, F., Chau, K.T., Chan, C.C., and Liu, C. (2013) Fault Diagnosis of Power Components in Electric Vehicles. *J. Asian Electr. Veh.*, **11** (2), 1659–1666.
6. Ma, S., Jiang, M., Tao, P., Song, C., Wu, J., Wang, J., Deng, T., and Shang, W. (2018) Temperature effect and thermal impact in lithium-ion batteries: A review. *Prog. Nat. Sci. Mater. Int.*, **28** (6), 653–666.
7. Buccolini, L., Ricci, A., Scavongelli, C., DeMaso-Gentile, G., Orcioni, S., and Conti, M. (2016) Battery Management System (BMS) simulation environment for electric vehicles. *2016 IEEE 16th Int. Conf. Environ. Electr. Eng.*, 1–6.
8. Wikner, E., and Thiringer, T. (2018) Extending battery lifetime by avoiding high SOC. *Appl. Sci.*, **8** (10).
9. Li, C., Cui, N., Wang, C., and Zhang, C. (2021) Simplified electrochemical lithium-ion battery model with variable solid-phase diffusion and parameter identification over wide temperature range. *J. Power Sources*, **497**, 229900.
10. Grolleau, S., Delaille, A., and Gualous, H. (2013) Predicting lithium-ion battery degradation for efficient design and management. *World Electr. Veh. J.*, **6** (3), 549–554.
11. Xu, J., and Cao, B. (2015) Battery Management System for Electric Drive Vehicles – Modeling, State Estimation and Balancing, in *New Applications of Electric Drives* (eds. Chomat, M.), InTechOpen.
12. He, H., Xiong, R., and Fan, J. (2011) Evaluation of Lithium-Ion Battery Equivalent Circuit Models for State of Charge Estimation by an Experimental Approach. *Energies* , **4** (4).
13. Johnson, V.H. (2002) Battery performance models in ADVISOR. *J. Power Sources*, **110** (2), 321–329.
14. Liu, X., Li, W., and Zhou, A. (2018) PNGV Equivalent Circuit Model and SOC Estimation Algorithm for Lithium Battery Pack Adopted in AGV Vehicle. *IEEE Access*, **6**, 23639–23647.
15. Gu, W.B., and Wang, C.Y. (2000) Thermal-Electrochemical Modeling of Battery Systems. *J. Electrochem. Soc.*, **147** (8), 2910.
16. Anwar, S., Zou, C., and Manzie, C. (2014) Distributed Thermal-Electrochemical Modeling of a Lithium-Ion Battery to Study the Effect of High Charging Rates. *IFAC Proc. Vol.*, **47** (3), 6258–6263.
17. Chang, W.-Y. (2013) The State of Charge Estimating Methods for Battery: A Review. *ISRN Appl. Math.*, **2013**, 953792.

18. Yatsui, M.W., and Bai, H. (2011) Kalman filter based state-of-charge estimation for lithium-ion batteries in hybrid electric vehicles using pulse charging. *2011 IEEE Veh. Power Propuls. Conf.*, 1–5.
19. Xu, L., Wang, J., and Chen, Q. (2012) Kalman filtering state of charge estimation for battery management system based on a stochastic fuzzy neural network battery model. *Energy Convers. Manag.*, **53** (1), 33–39.
20. Barbarisi, O., Vasca, F., and Glielmo, L. (2006) State of charge Kalman filter estimator for automotive batteries. *Control Eng. Pract.*, **14** (3), 267–275.
21. Zhang, J., and Xia, C. (2011) State-of-charge estimation of valve regulated lead acid battery based on multi-state Unscented Kalman Filter. *Int. J. Electr. Power Energy Syst.*, **33** (3), 472–476.
22. Tong, B., Wang, G., and Sun, X. (2014) Investigation of the Fluid-Solid Thermal Coupling for Rolling Bearing under Oil-Air Lubrication. *Adv. Mech. Eng.*, **7** (2), 835036.
23. Noura, N., Boulon, L., and Jemeï, S. (2020) A review of battery state of health estimation methods: Hybrid electric vehicle challenges. *World Electr. Veh. J.*, **11** (4), 1–20.
24. Shen, P., Ouyang, M., Lu, L., Li, J., and Feng, X. (2018) The Co-estimation of State of Charge, State of Health, and State of Function for Lithium-Ion Batteries in Electric Vehicles. *IEEE Trans. Veh. Technol.*, **67** (1), 92–103.
25. Kim, J., and Cho, B.H. (2011) State-of-Charge Estimation and State-of-Health Prediction of a Li-Ion Degraded Battery Based on an EKF Combined With a Per-Unit System. *IEEE Trans. Veh. Technol.*, **60** (9), 4249–4260.
26. Benbouzid, M.E.H. (2000) A review of induction motors signature analysis as a medium for faults detection. *IEEE Trans. Ind. Electron.*, **47** (5), 984–993.
27. Miljkovic, D. (2015) Brief review of motor current signature analysis. *HDKBRInfo Mag.*, **5**, 14–26.
28. Praneeth, A.V.J.S., and Williamson, S.S. (2017) Algorithm for prediction and control of induction motor stator interturn faults in electric vehicles. *2017 IEEE Transp. Electrif. Conf. Expo*, 130–134.
29. Fontes, A.S., Cardoso, C.A. V, and Oliveira, L.P.B. (2016) Comparison of techniques based on current signature analysis to fault detection and diagnosis in induction electrical motors. *2016 Electr. Eng. Conf.*, 74–79.
30. Thomson, W.T., and Culbert, I. (2017) Motor Current Signature Analysis for Induction Motors, in *Current Signature Analysis for Condition Monitoring of Cage Induction Motors: Industrial Application and Case Histories*, IEEE, pp. 1–37.
31. Singh, S., Kumar, A., and Kumar, N. (2014) Motor Current Signature Analysis for Bearing Fault Detection in Mechanical Systems. *Procedia Mater. Sci.*, **6**, 171–177.
32. Errabelli, R.R., and Mutschler, P. (2012) Fault-Tolerant Voltage Source Inverter for Permanent Magnet Drives. *IEEE Trans. Power Electron.*, **27** (2), 500–508.

33. Rodriguez, M.A., Claudio, A., Theilliol, D., and Vela, L.G. (2007) A New Fault Detection Technique for IGBT Based on Gate Voltage Monitoring. *2007 IEEE Power Electron. Spec. Conf.*, 1001–1005.
34. Fuchs, F.W. (2003) Some diagnosis methods for voltage source inverters in variable speed drives with induction machines - a survey. *IECON'03. 29th Annu. Conf. IEEE Ind. Electron. Soc.* (IEEE Cat. No.03CH37468), **2**, 1378-1385 Vol.2.
35. Ribeiro, R.L. de A., Jacobina, C.B., Silva, E.R.C. da, and Lima, A.M.N. (2003) Fault detection of open-switch damage in voltage-fed PWM motor drive systems. *IEEE Trans. Power Electron.*, **18** (2), 587–593.
36. Peuget, R., Courtine, S., and Rognon, J.-. (1998) Fault detection and isolation on a PWM inverter by knowledge-based model. *IEEE Trans. Ind. Appl.*, **34** (6), 1318–1326.
37. Jung, S., Park, J., Kim, H., Kim, H., and Youn, M. (2009) Simple switch open fault detection method of voltage source inverter. *2009 IEEE Energy Convers. Congr. Expo.*, 3175–3181.
38. Kim, S.Y., Nam, K., Song, H., and Kim, H. (2008) Fault Diagnosis of a ZVS DC–DC Converter Based on DC-Link Current Pulse Shapes. *IEEE Trans. Ind. Electron.*, **55** (3), 1491–1494.
39. Mamat, M.R., Rizon, M., and Khanniche, M.S. (2006) Fault Detection of 3-Phase VSI using Wavelet-Fuzzy Algorithm. *Am. J. Appl. Sci.*, **3** (1 SE-Research Article).

Index